Palgrave Studies in Law, Neuroscience, and Human Behavior

Series Editors
Marc Jonathan Blitz, Law, Oklahoma City University School of Law, Oklahoma City, OK, USA
Jan Christoph Bublitz, Faculty of Law, University of Hamburg, Hamburg, Hamburg, Germany
Jane Campbell Moriarty, Duquesne University School of Law, Pittsburgh, PA, USA

Neuroscience is drawing increasing attention from lawyers, judges, and policy-makers because it both illuminates and questions the myriad assumptions that law makes about human thought and behavior. Additionally, the technologies used in neuroscience may provide lawyers with new forms of evidence that arguably require regulation. Thus, both the technology and applications of neuroscience involve serious questions implicating the fields of ethics, law, science, and policy. Simultaneously, developments in empirical psychology are shedding scientific light on the patterns of human thought and behavior that are implicated in the legal system. The Palgrave Series on Law, Neuroscience, and Human Behavior provides a platform for these emerging areas of scholarship.

Stephan Schleim

Brain Development and the Law

Neurolaw in Theory and Practice

Stephan Schleim
Department for Theory and History
of Psychology
University of Groningen
Groningen, The Netherlands

ISSN 2946-5192   ISSN 2946-5206   (electronic)
Palgrave Studies in Law, Neuroscience, and Human Behavior
ISBN 978-3-031-72361-2   ISBN 978-3-031-72362-9   (eBook)
https://doi.org/10.1007/978-3-031-72362-9

Cover illustration: Prostock-studio/Alamy Stock Photo

This Palgrave Macmillan imprint is published by the registered company Springer Nature Switzerland AG
The registered company address is: Gewerbestrasse 11, 6330 Cham, Switzerland

If disposing of this product, please recycle the paper.

To all those who seek the solution to real-world problems in more than one discipline.

PREFACE

The challenge of pinpointing the fuzzy concept of maturity is hardly constrained to neuroscience. There is widespread lack of agreement on the age at which individuals should be considered adults (with the associate rights and protections) based on psychological indicators of maturity as well. (psychology professor Leah H. Somerville (Somerville, 2016, p. 1164))

It is true that open debate is an essential part of both legal and scientific analyses. Yet there are important differences between the quest for truth in the courtroom and the quest for truth in the laboratory. Scientific conclusions are subject to perpetual revision. Law, on the other hand, must resolve disputes finally and quickly. (Opinion of the Court, Daubert v. Merrell Dow Pharmaceuticals, Inc, 1993)

The interaction of neuroscience and law—"neurolaw" for short—has received increasing attention in recent years. In this book, the focus is on brain development and what this can mean for legal age limits and, in particular, criminal responsibility. My method is interdisciplinary, combining perspectives from psychology, law, cognitive and neuroscience, philosophy and history. Accordingly, brains and their activities are placed in temporal and spatial contexts in order to understand individual cognitive processes and answer normative questions.

This is in line with what is now known as the 4E view of cognition as embodied, embedded, enacted and extended: Our bodies are more than just nurturing shells of our brains, namely the anchor point of our perception and interaction with the world; this world continuously invites us to behave and we also make use of its tools to achieve certain goals. Ultimately, a living organism cannot be understood without its environment (e.g. Varela et.al., 2017).

We will address relevant historical examples since the 19th century. In Chapter 1, we will use biological psychiatry as a benchmark for the applications in neurolaw discussed later. Chapters 2 and 3 deal with important foundations of development, law and morality. Knowledge from psychology, brain research and criminal law is also provided. We will see more than once that drawing boundaries, both in science and in the field of law, is complex and often allows for multiple possibilities, especially for the period of adolescence (Somerville, 2016). But I hope that in the end you will agree with me that it was worth it: Because there are many essential things to learn here about us humans, science and our societies.

Groningen, The Netherlands   Stephan Schleim

References

Somerville, L. H. (2016). Searching for signatures of brain maturity: what are we searching for?. *Neuron*, 92(6), 1164–1167.

Varela, F. J., Thompson, E., & Rosch, E. (2017). *The Embodied Mind, Revised Edition: Cognitive science and human experience*. Cambridge, MA: MIT Press.

ACKNOWLEDGMENTS

First of all, I would like to thank the library of the University of Groningen for making this possible as an open access publication with a generous grant. Then, thanks to my colleagues in the Department of Theory and History of Psychology at the University of Groningen and in particular Dr. Jeremy T. Burman for helpful comments and a small financial contribution. I am similarly grateful for the discussion and comments in the Theoretical Psychology colloquium at the Sigmund Freud University, Vienna. I would also like to thank Dr. Fabian Hutmacher at the University of Würzburg for helpful comments on the first two chapters as well as two anonymous peer reviewers for their constructive feedback on the whole manuscript. In the almost 20 years that I have been working on this topic, I have been able to learn a lot from many more scholars—too many to mention here. Finally, my thanks also go to artificial intelligence, which has simplified translations; all its results have been carefully checked.

CONTENTS

List of Figures

LIST OF TABLES

CHAPTER 1

Introduction: Neuro, Psychiatry, Ethics and Law

> Ideologies, philosophies, religious doctrines, world-models, value systems, and the like will stand or fall depending on the kinds of answers that brain research eventually reveals. It all comes together in the brain. (Neurobiologist Roger W. Sperry (1913–1994) in the year he received the Nobel Prize for his research on split-brain patients [Sperry, 1981, p. 4])

At the beginning of a book or article like this one, reference is regularly made to the "Decade of the Brain." This is the title of a proclamation made by then US President George H. W. Bush (1924–2018) in 1990.[1] What is rarely explained is what this proclamation actually said: In short, it was a statement that although neuroscience has produced important knowledge, many challenges remain.

In particular, the treatment of neurogenetic and neurodegenerative diseases like Alzheimer's, stroke, schizophrenia and autism was mentioned as areas of application. Drug addiction was dealt with in a separate paragraph, with a reference to another proclamation by a former US president, the declaration of the "War on Drugs" from 1971. Or to put it simply, like Sperry, in one sentence: "It all comes together in the brain" (Sperry,

---

[1] https://www.govinfo.gov/content/pkg/STATUTE-104/pdf/STATUTE-104-Pg5324.pdf.

© The Author(s) 2025

S. Schleim, *Brain Development and the Law*, Palgrave Studies in Law, Neuroscience, and Human Behavior,
https://doi.org/10.1007/978-3-031-72362-9_1

1981, p. 4). There is another interesting connection to Sperry's essay entitled "Changing Priorities." The brain researcher and Nobel Prize winner called on his colleagues to align their research priorities with solving practical problems. The US president then formulated what those were in 1990 in his proclamation.

The statements about the new knowledge and the remaining challenges are probably no less true in 2025 than they were in 1990 and it is quite possible that they will be just as relevant in 2060. The extent to which neuroscience has influenced other disciplines has been explored elsewhere (Littlefield & Johnson, 2012; Pickersgill & Van Keulen, 2011; Schleim, 2014). Terms such as "neuropsychiatry," "neuropsychology" but also "neuroeconomics" or "neurotheology" testify to this process. This book is primarily concerned with the possible influence on the legal system. It focuses on the question of whether and how norms can be derived from brain development. Specific examples from various countries are addressed in Chapter 4.

Before we look at the role of the brain in law, there are some important basics to get across. It is one thing to announce major breakthroughs and changes. It is quite another whether they actually happen. As we shall see, major legal upheavals were announced as early as the nineteenth century on the basis of scientific—at that time: physiological—discoveries. The examples central to this book, however, arose following the "Decade of the Brain."

It goes without saying that the US president's declaration of 1990 did not simply fall from the sky, but reacted to and combined various earlier developments. For our purposes, psychiatry is of particular importance. This is because neuro-research has not only been particularly longstanding and strong in this field. Rather, as part of medicine, it is also a *practical* science, naturally related to application, and the forensic examples in later chapters also fall within its field.

In the next section, we will therefore look at how brain thinking (re-) entered psychiatry and gained the upper hand there. The 1980s are particularly important in this regard. We will then look at the new discourses of neuroethics and neurolaw since the early 2000s. The main part of the book then follows with Chapters 2–4, where important findings on the psychological and neuronal development of humans are first summarized. This is because the legal examples that follow in Chapter 4 largely revolve around brain development. In Chapter 5, I will offer my own pragmatic proposal on the relationship between behavior and the nervous system and finish the book with a general summary and outlook.

1.1   1980s AND 1990s: FROM THE "BROKEN BRAIN" TO THE "DECADE OF THE BRAIN"

The fact that psychological problems—think of anxiety, lack of attention or a persistently low mood—are often viewed as brain problems today can be seen in the increasing number of prescriptions for various psychotropic drugs. One example of this are the stimulants amphetamine and methylphenidate, which are prescribed as drugs called Adderall and Ritalin, among other brand names. This often happens after a diagnosis of attention deficit/hyperactivity disorder (ADHD).

In fact, many decades ago, such stimulants were used on a daily basis by many to reduce fatigue or improve mood (Rasmussen, 2008). They used to be available without a prescription, but since the "War on Drugs" they have been considered dangerous narcotics. Amphetamine was and is popular as a street drug, then named "speed," and for military purposes (Snelders & Pieters, 2011). And although the World Health Organization (WHO) already warned in the 1990s about the sharp increase in the administration of these drugs, particularly to children, prescriptions have multiplied since then (Schleim, 2023). Despite this decades-long increase, just as I write these words, another article appeared in the *New York Times* and claimed that ADHD in adults is still too often overlooked.[2]

This is just one example of what it can mean to understand mental health problems as brain problems: Then a "solution" with pharmacological means seems obvious. The theoretical foundations, individual and social consequences of this practice have already been discussed elsewhere (Davis, 2020; Szasz, 1974; Valenstein, 1998). For us, the precursors of the "Decade of the Brain" are relevant at this point.

The Brain as an Organ of the Psyche

It is often forgotten that mental disorders were already regarded as physical illnesses in ancient times. A clear testimony to this is a visit by

---

[2] https://www.nytimes.com/2024/05/20/well/mind/adhd-adults-diagnosis-treatment.html.

the famous physician Hippocrates (c. 460–c. 370 BC) to the materialist natural philosopher Democritus (c. 460–c. 370 BC). The latter suffered, as we would probably say today, from depression. When the doctor came to see him, he found dissected animals at his patient's home. The philosopher explained this as follows:

> How could I otherwise write on the nature of madness, its causes and the mode of alleviating it? The animals which thou seest here opened – I opened them not because of hate of the work of the divinity, but because I am searching for the seat and the nature of bile; for thou knowest it is usually, when it is excessive, the cause of madness. (quoted from Zilboorg & Henry, 1969, p. 45)

The idea that diseases are caused by an imbalance of humors in the body has influenced medicine for thousands of years, in both Western and Eastern cultures, in the north and south of the globe (Hall, 1971). The name "melancholia" (Gr. *mélas* = black and *cholé* = bile) bore witness to this until our recent past (Fig. 1.1).

On the way to what we now call "major depressive disorder," there was still the intermediate step of "melancholic depression" in the twentieth century (Shorter, 2015). Long before the discovery of neurotransmitters, it was assumed that depressed mood, for example, was caused by an excess of black bile. However, anatomical research in modern times revealed that this substance, which was thought to be found in the spleen, does not actually exist. Its alleged role was given to the neurotransmitter serotonin in the twentieth century, but then as an assumed deficiency instead of an excess; however, the serotonin hypothesis of depression is still controversial today (Cowen & Browning, 2015; Moncrieff et al., 2023; Valenstein, 1998). And who would have expected that the recent ICD-11 of the WHO reintroduces a "melancholic depression," such that the ancient "black bile disease" still casts its shadows into our times?

So if biological psychiatry presently assumes that mental disorders are physical disease entities, as in other parts of medicine, then this thinking is not new. Over time, however, the brain gained in importance as the central organ of psychiatry. Wilhelm Griesinger (1817–1868) is sometimes regarded as the "father of neuropsychiatry" because in the mid-nineteenth century he clearly described mental disorders as brain diseases (Schleim, 2022). However, such efforts can be traced back to

Fig. 1.1 Edvard Munch (1863–1944) who himself had severe psychological problems painted "Melancholy" in 1893. Today he might have called it "Depression." The example illustrates the historicity of the way we talk about mental states (License: public domain)

the seventeenth century (Berrios & Marková, 2002) and even antiquity (van der Eijk, 2005).

Speaking of melancholy, the English physician and pharmacist John Haslam (1764–1844) is another example worth mentioning. In 1809, he published the second edition of his *Observations on Madness and Melancholy* which described 37 case studies with brain examinations after the death of the patients. From these observations he concluded:

> From the preceding dissections of insane persons, it may be inferred, that madness has always been connected with disease of the brain and of its membranes. [...] It may be a matter, affording much diversity of opinion, whether these morbid appearances of the brain be the cause or the effect of madness: it may be observed that they have been found in all states of the disease. (Haslam, 1809, pp. 238–239)

This coincided with the emergence of phrenology, which linked psychological faculties to areas of the brain and externally recognizable characteristics of the head shape. Of course, the proponents of this doctrine also claimed that mental disorders were based on organic brain disorders; we will return to this briefly in Chapter 3. However, Haslam was not a phrenologist and cited the Irish philosopher John Toland (1670–1722) in support of his theory, who in his *Pantheisticon* of 1720 described all thinking as a function of the brain.[3]

In addition to discussing the question of whether the characteristics he found in the brain really are *the causes* of mental disorders, Haslam also made some other extraordinarily modern and topical observations. I will therefore briefly discuss two of them: For example, he argued that mania and melancholia should not be understood as separate disorders because, firstly, he had not been able to identify any brain differences between them and, secondly, the treatment was the same. He refuted the alternative hypothesis that these are not *physical* but rather *mental* diseases by pointing out that material medical procedures—but not logical-rational methods—worked.

We do not know exactly what John Haslam thought he saw in the brains of these patients around 1800; he described tissues as abnormally soft or firm (for illustrative purposes: Fig. 1.2). But we do not need to establish a history of psychiatry or medical psychology here, as the discipline was once called (Horwitz, 2020; Zilboorg & Henry, 1969). For our purposes, it is relevant that the search for "the psyche" in the brain is still ongoing today. We should keep these historical examples in mind when we now turn our attention to the new waves of neuro- or biological psychiatry in the twentieth century. Particularly when we look at the allegedly revolutionary idea of "broken brains" in the 1980s, we will understand the historical background of this thinking.

"Broken Brains"

Psychiatry had many faces in the twentieth century. Under the influence of Sigmund Freud (1856–1939), the psychodynamic approach led the way for several decades. According to this view, mental disorders

---

[3] "Cogitatio [...] est inotus peculiaris Cerebri, quod hujus facultatis est proprium organum", quoted from Haslam (1809, p. 240). "Thought [...] is a peculiar feature of the brain, which is the proper organ of this faculty".

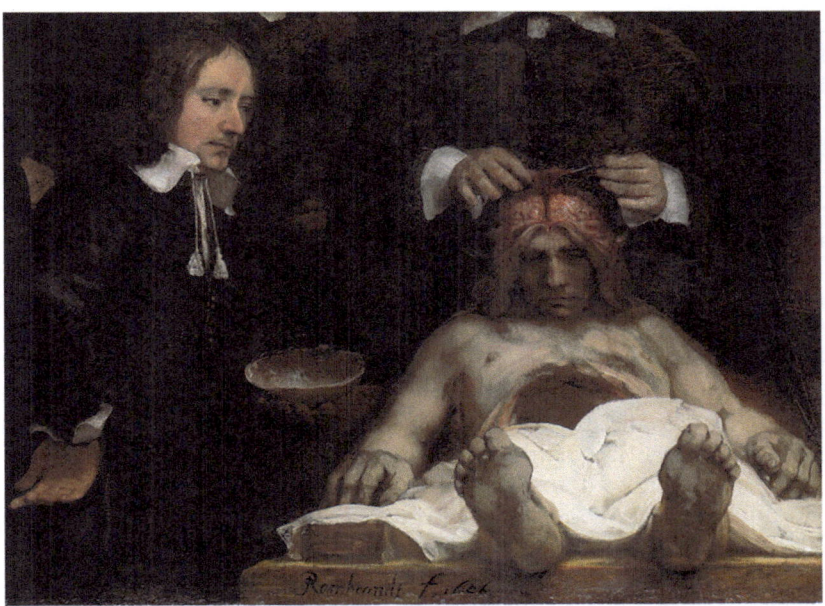

Fig. 1.2 Haslam's sections may have resembled "The Anatomy Lesson of Dr. Deijman" (1656), which the Dutch master painter Rembrandt van Rijn (1606/1607–1669) captured on this canvas. However, only this fragment survived a fire in 1723. On the dissection table lies the body of Joris Fontein, who died in his early 20s: he was caught in the act of burglary, confessed (under torture) to numerous other crimes and was sentenced to death by hanging. After his execution, his body was given to the Amsterdam Surgeons' Guild. We do not know whether the dissecting doctor searched the dead man's brain for the cause of the crime; if so, it would be an early example of neurolaw. However, surgeons recently determined from a replication of the dissection that the pose depicted presupposes a broken neck, as can occur during hanging (IJpma et al., 2013) (*Source* Amsterdam Museum. License: public domain)

often arise from unconscious conflicts in the "psyche" that are frequently associated with early childhood experiences. The disorders are then to be treated primarily with talk therapy, for example with psychoanalysis, to make the unconscious dynamics conscious and thus resolve the conflicts. The American diagnostic manuals DSM-I from 1952 and DSM-II from 1968 were strongly influenced by this kind of thinking (Mayes & Horwitz, 2005).

In the 1930s–1970s, however, there were major breakthroughs in the field of biological or neuropsychiatry: think of the use of brain surgery, brain stimulation and pharmacology, for example lithium and chlorpromazine. At the time, these were even applied to criminal law problems, as an article in *Popular Science Monthly* shows (Fig. 1.3).

However, the high expectations turned out to be exaggerated time and again and—sometimes very severe—side effects eventually received more attention (Delgado, 1971; Schleim, 2021; Valenstein, 1973, 1998; Williams et al., 2008). The lack of more objective criteria for diagnosis, such as blood, genetic or brain tests, led to confusion. For example, schizophrenia was diagnosed more frequently in New York, but depression more frequently in London. Finally, studies with standardized material showed that psychiatrists in the USA and Great Britain had different ideas about these disorders (Kendell et al., 1971; Schleim, 2023).

With advances in genetics, imaging techniques and the information sciences, which we now summarize as "neuroscience," the dream of a modern scientific psychiatry seemed within reach. For the DSM-III of 1980, a group of psychiatrists who found the old psychodynamic categories too speculative and who wanted to adapt the discipline to scientific advances finally prevailed (Mayes & Horwitz, 2005; Wilson, 1993). This thinking still characterizes the DSM today. But this advance also implied the removal of the etiology, the theory of causes, from the diagnostic manual and working groups at the conference table agreed on the now well-known checklists for several hundred mental disorders.

Until the much-delayed DSM-5 was published in 2013, one of the main objectives was to provide the neurobiological etiology that was missing from the new approach (Hyman, 2007). At the turn of the millennium, much was expected from genetics and brain imaging research in particular. For example, studies such as the one by Ingvar and Franzén had already reported differences in the blood flow of the frontal brain of people diagnosed with schizophrenia in the 1970s (Ingvar & Franzén, 1974). Prior to this, over 100 years of anatomical brain examinations of deceased patients with such severe disorders had not provided any clarity. Proponents of this approach, such as the German-Austrian neuroanatomist Theodor Meynert (1833–1892), who localized mental

Fig. 1.3 "Have You a Wrong Way Brain?" asked the headline of an article in the July 1939 issue of *Popular Science Monthly*. Inspired by brain surgery, the author speculated that criminal behavior occurs when the less dominant hemisphere of the brain takes control and "Dr. Jekyll becomes Mr. Hyde" or a loving father becomes a villain. He also wrote that up to 85 percent of prisoners suffer from mental disorders and procedures such as lobotomy or insulin shock therapy—both no longer used today—could solve the problems. The concluding sentence was: "By cutting the roots of crime in the minds of malefactors, they may some day play a major role in reducing our $15,000,000,000-a-year crime bill and in turning outlaws into good citizens" (*Source*: created with Adobe Firefly)

disorders only in the frontal brain (Meynert, 1884), were even accused of practicing "brain mythology" by psychiatrists focusing on actual patients in the clinics (Marx, 1970).

In the twentieth century, interest in this approach to psychiatry continued in waves (Fig. 1.4). Thanks to new scientific procedures and the new categories of the DSM-III, the 1970s and 1980s were characterized by a spirit of optimism. We already saw this in Roger Sperry's far-reaching statement at the beginning of this chapter (Sperry, 1981) and it is also very well illustrated in Nancy Coover Andreasen's much-cited book *The Broken Brain: The Biological Revolution in Psychiatry* from 1984. The American psychiatrist became one of the leading experts in the field of schizophrenia, helping to shape both the DSM-III of 1980 and the DSM-IV of 1994, and for 13 years was editor-in-chief of the *American Journal of Psychiatry*. In her book, she announced a revolution in research, diagnosis and therapy through the biological model. In her words:

> In more recent times the biological model has been shaped by the growth of the discipline 'neuroscience' or the neurosciences. [...] Much of the time these abnormalities [in behavior, emotions, and thinking, St. S.] cannot be traced to a distinct area of damage in the brain, although the biological model assumes that as our knowledge progresses, some type of malfunction in the brain will be found. The current biological revolution in psychiatry places great emphasis on the search for the physical causes of mental illness. (Andreasen, 1984, pp. 27–28)

The goal was therefore clear: to find the dysfunctions in the nervous system at a cellular or molecular level that cause mental health problems. In addition to electroencephalography (EEG), which had already been developed in the 1930s, new imaging techniques such as computer tomography (CT), positron emission tomography (PET) and functional magnetic resonance imaging (fMRI)—which was only developed a few years after Andreasen's book—were to provide deeper insights into the living human brain than ever before. With the help of genetic, pharmacological and electrical tests in animals and humans, the underlying mechanisms should be identified. Andreasen explains this with a case study, which I would like to reflect upon below. But first, some basic knowledge about how imaging techniques work.

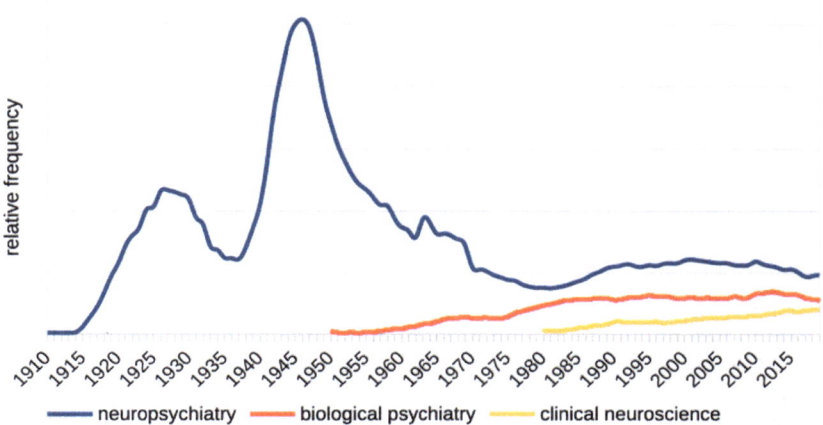

Fig. 1.4 This Google Ngram shows the relative frequency of the terms "neuropsychiatry," "biological psychiatry" and "clinical neuroscience" in English-language books. The large spike in the 1940s and 1950s coincides with the spread of psychosurgery and new psychotropic drugs. Later, other terms also appear more frequently. However, it should be borne in mind that in the course of the twentieth century, publications in books became less important in research and were increasingly replaced by articles in specialist journals (scale: 10^{-8} percent) (*Source* Google Ngram)

Understanding Neuroimaging

Imaging techniques have become an integral part of medicine, for example in the search for tumors or the examination of the fetus during pregnancy. But their importance for research has also steadily increased. Today, brain scans are one of the most important measurement methods not only in neuroscience, but also in psychology and psychiatry (Friston, 2009). The distinction between *structural* and *functional* methods is important.

Structural procedures make different types of tissue or bone visible. The best known is probably still X-ray radiation, which is absorbed to different degrees by different types of tissue and thus leads to visible differences on the images. Certain—often radioactive—tracer substances, which are administered intravenously, can be used to highlight certain types of tissue, such as tumors.

In the neurosciences, functional methods are used to examine brain function instead. In the past, tracer substances were also used for this purpose, for example to visualize the consumption of the energy supplier glucose or the presence of certain synapses responding to neurotransmitters. The great advantage of functional magnetic resonance imaging (fMRI) is that it does not require tracer substances. The physiological basis for this was discovered around 1990 by the Japanese biophysicist Seiji Ogawa and his colleagues. This is based on the fact that blood has different magnetic properties depending on its oxygen saturation, which can be measured in strong magnetic fields; and the oxygen saturation is in turn associated with neuronal activity (Logothetis, 2008). However, this correlation is not perfect and oxygen saturation is only an indirect but often useful indicator of neuronal activity.

For the purposes of this book, two observations are important: First, the known brain images reflect reactions in experimental situations that can only be interpreted by comparing different conditions—often a target and control condition. Second, the colors projected onto an anatomical brain image are not a direct visualization of brain activity, but of the results of statistical tests (Dumit, 2004; Schleim & Roiser, 2009).

This means that the results of fMRI studies are dependent on the characteristics of the experiment and the evaluation by the researchers. They are therefore by no means as direct or objective representations of experience and thought as is often portrayed (Racine et al., 2010). In fact, it has now been shown that different neuroscientists can draw different conclusions from the same brain data (Botvinik-Nezer et al., 2020).

People in Contexts

At the beginning of her book about *Broken Brains* and the biological revolution in psychiatry, Andreasen referred to the problematic past and present of her discipline. In the past, people with mental health problems were persecuted as being possessed by the devil or evil spirits and, in extreme cases, even killed. The psychiatrist did not mention that they were sometimes regarded as saints. She then described the case of the American sailor William (actually James) Norris, who was brought to London's Bethlem Royal Hospital as a "lunatic" in 1800. Due to violent outbursts, he was eventually put in chains and kept in this state of severely restricted freedom of movement for around ten years.

During a parliamentary inquiry into the hospital, the appalling conditions came to light and were disseminated to the public as horror stories. A parliamentary commission set up in 1815 then led to an improvement in the conditions in British hospitals and "lunatic asylums." Although Norris was freed from his chains, he died of tuberculosis shortly afterward—probably also due to his poor physical condition as a result of his long imprisonment.

The case was so important to Andreasen that she placed a drawing of the chained patient at the beginning of her book. According to historical reports, however, Norris was also considered manipulative and dangerous, almost killing his guard and biting off another patient's finger (Andrews et al., 1997). To alleviate his fate, he was given a cat, newspapers and books. Andreasen did not mention any of these circumstances. She also seems to have been unaware that the doctor blamed for the scandal was none other than John Haslam, the "brain doctor" we met above, thus a medical professional who wanted to advance psychiatry in the same way she did. Unfortunately, violent patients still pose practical and legal challenges for psychiatric institutions today.

Why such cases are important for our book becomes particularly clear from the portrayal of "Bill," to whom Andreasen devoted almost eight whole pages: This psychiatric patient had studied medicine at Harvard with outstanding achievements. Toward the end of his studies, his father died. Bill's subsequent depression was treated with medication and slowly improved. However, due to his psychological problems, the examination board initially denied him his degree and required him to study for a year longer. He kept his feelings of humiliation and despair to himself.

Despite his exceptional academic achievements, he then had difficulty finding the necessary internships and a position as a junior doctor. The stigma of "psychiatric illness" clung to him. He eventually married a former fellow student, had two daughters with her and set up his own practice. He had another depressive episode treated by a psychiatrist in a community 50 miles away, as he was aware of the stigma of this diagnosis. The new drug therapy helped him.

When Bill was 35 years old, his wife was diagnosed with incurable liver cancer. She died within a year. During the last two months of her life, Bill was overcome by uncontrollable crying fits at work. Eventually, he himself was admitted to a psychiatric hospital. His terminally ill wife resented this, and his acquaintances also had little understanding for it: "Some people in the community had trouble understanding why he could not maintain better control of himself. They thought him weak and lacking in self-discipline" (Andreasen, 1984, p. 4).

After the funeral, Bill threw himself into work. He found a house-keeper to look after his daughters, who were two and four years old when his wife died. After a year, he looked for a new partner and eventually married Joann, ten years his junior. While he longed for a domestic life, she preferred to go out a lot and eventually had an open relationship. Bill could not stand her affairs with other men in the long term. When he asked her to be monogamous, she divorced him. This was the third time Bill had lost an important person: his father and his first wife through death, his second wife through divorce.

The fact that Joann stayed in the area and that he kept seeing her with new partners bothered him. After one such encounter, he got drunk and drove to her apartment in the evening. When he harassed her, Joann called the police. The local media reported on Bill's problem behavior. When he was again receiving medical treatment for depressive symptoms and could no longer make it to work every day, the public scandal was the last straw. According to Andreasen, he had no support:

> After his arrest and its bad publicity, he went to pieces. Other physicians in the community knew that he had been having problems, but no one made any attempt to offer help or sympathy. Some thought he simply needed to be tougher, while others thought he was behaving irresponsibly. Having put most of his energy into his work and his family, Bill had few close friends. The two or three whom he did have held back from approaching him because of uncertainty or embarrassment. (ibid., p. 6)

In the end, he was no longer able to work and was admitted to hospital. This time, no medication helped, so they tried electroconvulsive therapy (also known as electroshock therapy). Thanks to this, he felt better again and was able to leave the clinic after six weeks—only to find that his license to practice medicine had been temporarily revoked due to his stay in the clinic. Not only did he lose his daily job as a doctor and source of income, but he was once again the subject of the press, this time even on television. His daughters, who were now at school, were also drawn into the scandal.

Bill finally took his own life at the age of 45 so that, Andreasen wrote, he could at least leave his daughters the substantial life insurance payout. Although he got his license to practice medicine back, his depression did not disappear. Lonely and isolated, he probably felt like a failure. If he had gone back into hospital because of his mental health problems, he might have lost his license forever.

Not Just the Brain

Bill's case not only shows how complex and multi-layered a human life can be. What is interesting for us now are the conclusions that Nancy Coover Andreasen drew from it for the biological revolution in psychiatry:

> Psychiatry, like the prodigal son, has returned home to its place as a specialty within the field of medicine. It has become increasingly scientific and biological in its orientation. Psychiatry now recognizes that the serious mental illnesses are diseases in the same sense that cancer or high blood pressure are diseases. Mental illnesses are diseases that affect the brain, which is an organ of the body just as the heart or stomach is. People who suffer from mental illness suffer from a sick or broken brain, not from weak will, laziness, bad character, or bad upbringing. (Andreasen, 1984, p. 8)

According to the influential psychiatrist, the persecution of the "possessed," the incarceration of patients such as James Norris or ultimately Bill's suicide would never have occurred if society had only correctly understood the nature of psychological-psychiatric disorders, namely as *brain diseases*. From this perspective, they would be seen in the same way as cancers: without shame, guilt and stigma. Then patients would be treated with more compassion, understanding and patience. More than 30 years later, Kenneth S. Kendler of Virginia Commonwealth University would argue that psychiatric disorders and psychiatry can only be taken seriously with a biological foundation (Kendler, 2016). Like Andreasen, he is an influential psychiatrist who sought to discover the biological basis of schizophrenia, primarily through genetic research, and helped shape some editions of the DSM.

From today's perspective, exactly 40 years later, we can look back on these far-reaching expectations, of which Andreasen was a relevant example. During these decades, Steven E. Hyman (1996–2001), Thomas R. Insel (2002–2015) and Joshua A. Gordon (since 2016), thus three neuropsychiatrists, were directors of the world-leading US National Institute of Mental Health (NIMH). Its director now decides on an annual budget of over 2.5 billion dollars, most of which is invested in research.[4] Gordon even described his field as "circuit psychiatry" when he was appointed, referring to neuronal circuits allegedly underlying mental disorders (Gordon, 2016). Rodents such as mice and rats are therefore an important model organism for psychiatric research. The perspective of human patients is scarce, just like with Meynert's approach in the nineteenth century.

When the DSM-5 was published in 2013, the disappointment was great: *Not a single* reliable diagnostic biomarker had been found for any of the several hundred mental disorders differentiated therein. Clinical psychologists and psychiatrists commonly still have to talk to their patients in order to make a diagnosis, and the mode of action of the frequently prescribed psychotropic drugs, for example for depression, is still very controversial (Davies et al., 2023; Moncrieff et al., 2023; Szasz, 1974; Valenstein, 1998). Experts from various disciplines have dealt intensively with this problem and possible reforms of psychiatry (Frances, 2013; Fuchs, 2018; Rose & Rose, 2023; Schleim, 2023; Scull, 2021). Thomas Insel commented on the progress made after his tenure as NIMH Director as follows:

> I spent 13 years at NIMH really pushing on the neuroscience and genetics of mental disorders, and when I look back on that I realize that while I think I succeeded at getting lots of really cool papers published by cool scientists at fairly large costs – I think $20 billion – I don't think we moved the needle in reducing suicide, reducing hospitalizations, improving recovery for the tens of millions of people who have mental illness.[5]

Like many others in the meantime, Insel now recommends a stronger focus on the prevention of mental disorders and on the social and institutional side of treatment (Insel, 2022). A decade earlier, he himself had explained to the general public that the disorders were caused by "faulty circuits" and had announced that "[n]euroscience is revealing the malfunctioning connections underlying psychological disorders and forcing psychiatrists to rethink the causes of mental illness" (Insel, 2010, p. 44). In the case of depression, for example, a certain brain region would

---

[4] https://www.nimh.nih.gov/about/budget.

[5] https://www.wired.com/2017/05/star-neuroscientist-tom-insel-leaves-google-spawned-verily-startup/.

have to be "rebooted," as in the case of a crashed computer. Even if, to paraphrase Sperry, many things come together in the brain (Sperry, 1981), this central organ alone has not yet proven to be sufficient for understanding, let alone solving people's mental problems.

Psyche and Society

This brings us to the missing ingredient in understanding psychological-psychiatric disorders: Factors such as poverty, stress, relationship status, living in a city, severe life events, workplace organization, and the availability of help all play an important role in our mental health (Arango et al., 2021; Mirowsky & Ross, 2003; OECD, 2011; Rugulies et al, 2023; Sheldon et al, 2021; WHO, 2004). If we take another look at Bill's life from this perspective, the problem may look quite different:

In his biography, a combination of outstanding achievements and high functional pressure stands out. When his father died, he wanted (or was supposed) to resume and complete his studies after just one month. Having to repeat the last year was a humiliation for him. The extent to which grief after the loss of a loved one is normal or a characteristic of a depressive disorder has long been disputed (Frances, 2013). In the past, the therapist or psychiatrist played a decisive role in drawing these boundaries: They were supposed to assess the extent to which the grief conformed to cultural norms.

The DSM-5-TR of 2022 recently added "prolonged grief disorder" as a new category. Intense loneliness, experienced meaninglessness of life, intense emotional pain and dysfunction in everyday life are exemplary symptoms (APA, 2022). A mourning period of one year is considered normal for adults and six months for children and adolescents. Bill had only taken a fraction of this time for himself. Even after the death of his wife and the divorce, the pressure to function, this time as a doctor with his own practice, had always remained very high.

Another striking feature is the strong social rejection of mental health problems or their stigmatization in his environment. In part, this could be seen as the flip side of the pressure to function: Bill had no choice but to continue studying or working quickly because that was the only thing that was considered acceptable. He also didn't seem to be able to talk to friends or family about his feelings and fears. This was perhaps all linked to a certain ideal of masculinity that was particularly strong for the socially respected but also responsible role of a doctor. Andreasen explicitly described how he was expected to have more self-control and resilience.

Even decades later, the significantly higher suicide rate among men was explained by certain ideas of masculinity (Swami et al., 2008). The tendency of members of this gender to be unable to talk about their feelings (Chandler, 2022) is cited as an important factor, as is a certain understanding of heroism (Rasmussen et al., 2018). We remember the idea of sacrificing oneself so that one's daughters receive the high life insurance premium. Such psychosocial explanations were not considered for a long time because many researchers believed that the large gender difference in suicide rates *had to* be explained genetically (Swami et al., 2008).

The holistic view adopted by myself and others does not deny that Bill was also a biological being with a certain genetic disposition, a nervous system and a brain. However, considerable doubts are raised as to whether the problem—and therefore the solution—can be found primarily at this level. Even though psychotropic drugs and later the stimulation of his brain with a strong electrical current improved the symptoms, these interventions did nothing to change the inner and outer harshness and rejection that Bill had to deal with after the heavy blows of fate. In psychotherapy, the internalized thought patterns could have been questioned and thus perhaps gradually changed; and according to the social psychiatric approach, he could perhaps have been provided with an assistant for the practice or found an environment with less functional pressure.

Andreasen ends the introduction to her book with the remark that a better—by which she meant: biological—understanding of mental disorders unfortunately came too late for Bill (Andreasen, 1984). That her view was too one-sided is shown not only by the my contextualization of the case, but also by the sobering neuroscientific research results of the following 40 years.

A couple of years ago, I was invited to speak at the Berlin Psychiatry Days to talk about "the disappearance of the social from psychiatry." That was the only time in my career to date that I received a second round of applause "for such an important contribution" in front of hundreds of professionals, including many psychotherapists, while I had the impression of telling these clinicians with much more practical experience the most obvious things.

At the conference dinner, a leading biological psychiatrist sat next to me and said that research has *always* included the social: namely in epigenetics. This refers to the process by which genes are controlled by environmental influences. In my opinion, however, it illustrates the limited view of this school of thought that the significance of the social can only be imagined at the biological level, in terms of molecules activating or deactivating genes.

Practical Use

Why is this important for this book? Once again, there is no denying that our brains and bodies play a central role in our perception, experience, thinking and behavior. However, particularly in an applied context such as that of brain research and law, it is an important question what can be done with this knowledge in practice. Biological psychiatry serves as a comparison here: If, even after some 200 years of intensive research in this field, mental disorders cannot be generally classified, diagnosed, treated or at least the success of treatment documented on the basis of neurobiological examinations, then the assumed brain-psyche connection is of much less use for practical purposes than often claimed.

I'm not denying that psychotropic substances or electricity can be used successfully in many cases to manage undesired experiences, like that of anxiety or psychoses. Many people do the same without medical supervision to cope with the challenges in their lives: for example, soldiers in war areas, homeless people or students and athletes in competitive environments. I take the stance that it is mostly a social-normative decision to call one group "patients" and the other "drug (ab)users" (Schleim, 2023). The American opioid crisis illustrates how people can shift back and forth between these categories, depending on lawmakers' choices (Pieters, 2023). But according to my view, such experiences must also be seen in a psychosocial context and are not just localizable entities in brains, in contrast to what leading biological psychiatrists like Griesinger, Meynert, Andreasen, Insel or Gordon stated.

From the point of view of embodiment, it is trivially true that psychological processes can be influenced in the body, for in us humans they necessarily *are* embodied. The still frequently reiterated mind–body dualism became the dominant view in Christian theology—and was inherited as a reified separate "mind-thing" even by non-theistic philosophers, psychologists and neuroscientists (Schleim, 2020a). In contrast, it was self-evident for many philosophers and physicians of antiquity to view "the soul" as physiological.

For example, Rufus of Ephesus (c. 70–110 CE), for centuries the leading authority on melancholic depression, suggested wine, sex and hot baths to treat the disorder (Pormann, 2008). Now we know that the first affects the neurotransmitters GABA and glutamate in the brain, the second dopamine and the third the parasympathetic relaxation system. (This is a simplified account, but sufficient for our purposes.) However, also here the psychosocial context is important: What Rufus called "melancholy" arose among males in the socially demanding Greek aristocracy. We should not take for granted that such terms have the same meaning some 2000 years later, but a deeper analysis goes beyond the purview of this book.

Particularly when such basic conceptual and philosophical issues are unresolved, we should be guided by practical utility. Because the view of biological psychiatry turned out to be oversimplified, we should, by analogy, be careful and cautious in such a young field as neurolaw. These different perspectives—biology, psychology, society—will accompany us throughout the book. But let us now take a closer look at the topics falling within the scope of neuroscience, ethics and law.

1.2   2000s: Neuroethics and Neurolaw

We have taken a historical approach to neuro topics and have so far mainly focused on psychiatry. We realized that the high expectations for the study of the body, the nervous system and the brain go back further than the nineteenth century. In the 1970s and 1980s, there was a new wave of enthusiasm, fueled by advances in genetics and imaging techniques. The 1990s became the "Decade of the Brain." We have now reached the year 2000. From this time onwards, the ethical and legal aspects of neuroscience came increasingly into focus (Fig. 1.5).

"Neuroethics" is often understood as the complementary pair of ethics of neuroscience and neuroscience of ethics. The former could be seen as a continuation of long-established medical and bioethics. In fact, there were initially some critical voices that questioned the introduction of ever new ethics fields—for bioethics, gene ethics, neuroethics, nanoethics and so on. The special significance of the nervous system and brain for us humans was cited as an argument in favor of independent neuroethics (Roskies, 2007). As early as 2001, the journal *NeuroRehabilitation* published a

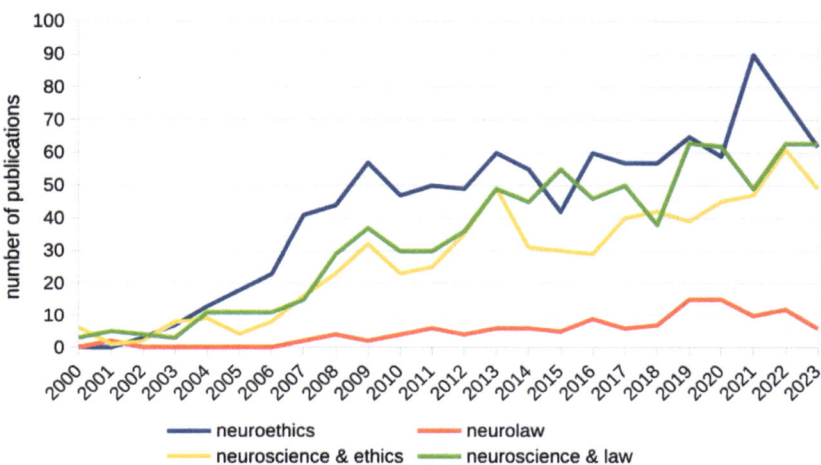

Fig. 1.5 Since 2000, the number of publications on the significance of neuro-science for ethics and law has been steadily increasing (*Source* Web of Science [topic search])

special issue on neurolaw (Tire, 2001). This was followed three years later by one in the *Philosophical Transactions of the Royal Society of London* (Zeki & Goodenough, 2004).

In 2006, the International Neuroethics Society was founded, which explicitly includes legal issues relating to advances in neuroscience.[6] Incidentally, its first president was the neuropsychiatrist Steven Hyman, whom we already know. One year later, the *American Journal of Bioethics-Neuroscience* was founded as the publication organ of this association, from 2010 as an independent journal. The journal *Neuroethics* is published since 2008. Topics such as neuroimaging, neuropsychopharma-cology, brain-computer interfaces and brain doping are now part of the core area of the discipline (Buniak et al., 2014; Schleim, 2020b).

In contrast, the neuroscience of ethics should provide new insights into the neuronal foundations of our moral decisions, as a kind of biological offshoot of the moral psychology that has been practiced for some time (Schleim, 2008, 2015). In fact, this was the research area of my own doctoral thesis, for which I was probably the first researcher in the world

[6] https://www.neuroethicssociety.org/about.

to systematically investigate lawyers in an fMRI scanner (Schleim et al., 2011). The debate about whether certain ethical conclusions can or even must be drawn from such empirical insights continues to this day (Racine et al., 2017). However, this discussion is a broad field and goes beyond the scope of this book.

Similarly, some describe "neurolaw" as a combination of the law of neuroscience and the neuroscience of law (Chandler, 2018). The former includes, for example, the regulation of neurotechnologies and the handling of brain data. The latter refers, among other topics, to how findings from brain research influence legal concepts or what they tell us about the behavior of decision-makers in the legal system. In fact, both directions play a role in this book, but the main focus is on the connection between brain development and the concept of responsibility (Chapter 4).

We will take a closer look at the other key topics in neurolaw later on. For now, we have become familiar with all the aspects that are important for an understanding of the topic and a preview of the rest of the book. After a brief summary of this introduction, the next chapter will take a closer look at the psychological and neurobiological development of humans.

1.3   Summary

Neuroscience was not only fascinating during the "Decade of the Brain," but also before and after it. After all, the brain, with its approximately 86 billion nerve cells and many more connections, is sometimes described as the most complex object known to us. In this introduction, however, we saw from the example of psychiatry in particular that it is difficult to apply this to individual cases in practice. Law, just like medical diagnoses and treatments, is often about individuals and their actions and rights. Even if we assume that our perceptions, experiences, thoughts and actions are embodied and in this sense also biological, this does not automatically guarantee the applicability of neuroscientific findings in practical contexts.

In this chapter, we saw that this type of research has a centuries-old history and that case studies should be seen in their psychosocial perspective. By taking a different stance, as with Bill, one arrives at different descriptions of problems and possible solutions. Nancy Andreasen's ideas about *Broken Brains* and the disappearance of the stigma of mental disorders has turned out to be too optimistic. Moreover, even today a psychiatric diagnosis can still lead to social exclusion (Franz et al., 2023;

Mann & Contrada, 2022; O'Connor & Joffe, 2013). In the worst case, the stigma becomes even greater through a neurobiological perspective if the problems are seen as particularly permanent or dangerous, as if they were "hardwired" in the brain.

Since the turn of the millennium, academics from various disciplines have been increasingly concerned with the ethical and legal challenges of neuroscience. This book also falls into this area. In the next chapter, we first look at the psychological and neuronal aspects of human development. We will then return to neurolaw in more detail and use various examples to analyze exactly how brain development and the law fit together.

REFERENCES

Andreasen, N. C. (1984). *The broken brain: The biological revolution in psychiatry.* Harper & Row.

Andrews, J., Briggs, A., Porter, R., Tucker, P., & Waddington, K. (1997). *The history of Bethlem.* Routledge.

APA [American Psychiatric Association]. (2022). *Diagnostic and Statistical Manual of Mental Disorders, fifth edition, Text Revision (DSM-5-TR).* American Psychiatric Association Publishing.

Arango, C., Dragioti, E., Solmi, M., Cortese, S., Domschke, K., Murray, R. M., et al. (2021). Risk and protective factors for mental disorders beyond genetics: An evidence-based atlas. *World Psychiatry, 20*(3), 417–436.

Berrios, G. E., & Marková, I. S. (2002). The concept of neuropsychiatry: A historical overview. *Journal of Psychosomatic Research, 53*(2), 629–638.

Botvinik-Nezer, R., Holzmeister, F., Camerer, C. F., Dreber, A., Huber, J., Johannesson, M., et al. (2020). Variability in the analysis of a single neuroimaging dataset by many teams. *Nature, 582*(7810), 84–88.

Buniak, L., Darragh, M., & Giordano, J. (2014). A four-part working bibliography of neuroethics: Part 1: Overview and reviews–defining and describing the field and its practices. *Philosophy, Ethics, and Humanities in Medicine, 9*, 1–14.

Chandler, A. (2022). Masculinities and suicide: Unsettling 'talk' as a response to suicide in men. *Critical Public Health, 32*(4), 499–508.

Chandler, J. A. (2018). Neurolaw and neuroethics. *Cambridge Quarterly of Healthcare Ethics, 27*(4), 590–598.

Cowen, P. J., & Browning, M. (2015). What has serotonin to do with depression? *World Psychiatry, 14*(2), 158.

Davies, J., Read, J., Kruger, D., Crisp, N., Lamb, N., Dixon, M., et al. (2023). Politicians, experts, and patient representatives call for the UK government to reverse the rate of antidepressant prescribing. *BMJ, 383.* https://doi.org/10.1136/bmj.p2730

Davis, J. E. (2020). *Chemically imbalanced: Everyday suffering, medication, and our troubled quest for self-mastery.* University of Chicago Press.

Delgado, J. M. R. (1971). *Physical control of the mind: Toward a psychocivilized society.* Harper & Row.

Dumit, J. (2004). *Picturing personhood: Brain scans and biomedical identity.* Princeton University Press.

Frances, A. (2013). *Saving normal: An insider's revolt against out-of-control psychiatric diagnosis, DSM-5, Big Pharma, and the medicalization of ordinary life.* Morrow.

Franz, D. J., Richter, T., Lenhard, W., Marx, P., Stein, R., & Ratz, C. (2023). The influence of diagnostic labels on the evaluation of students: A multilevel meta-analysis. *Educational Psychology Review, 35*(1), 17.

Friston, K. J. (2009). Modalities, modes, and models in functional neuroimaging. *Science, 326*(5951), 399–403.

Fuchs, T. (2018). *Ecology of the brain: The phenomenology and biology of the embodied mind.* Oxford University Press.

Gordon, J. A. (2016). On being a circuit psychiatrist. *Nature Neuroscience, 19*(11), 1385–1386.

Hall, T. S. (1971). Life, death and the radical moisture: A study of thematic pattern in medieval medical theory. *Clio Medica, 6,* 3–23.

Haslam, J. (1809). *Observations on madness and melancholy: Including practical remarks on those diseases, together with cases, and an account of the morbid appearances on dissection.* J. Callow.

Horwitz, A. V. (2020). *Between sanity and madness: Mental illness from ancient Greece to the neuroscientific era.* Oxford University Press.

Hyman, S. E. (2007). Can neuroscience be integrated into the DSM-V? *Nature Reviews Neuroscience, 8*(9), 725–732.

IJpma, F. F. A., Middelkoop, N. E., & van Gulik, T. M. (2013). Rembrandt's anatomy lesson of Dr. Deijman of 1656 dissected. *Neurosurgery, 73*(3), 381–385.

Ingvar, D., & Franzén, G. (1974). Distribution of cerebral activity in chronic schizophrenia. *The Lancet, 304*(7895), 1484–1486.

Insel, T. R. (2010). Faulty circuits. *Scientific American, 302*(4), 44–52.

Insel, T. R. (2022). *Healing: Our path from mental illness to mental health.* Penguin Press.

Kendell, R. E., Cooper, J. E., Gourlay, A. J., Copeland, J. R. M., Sharpe, L., & Gurland, B. J. (1971). Diagnostic criteria of American and British psychiatrists. *Archives of General Psychiatry, 25,* 123–130.

Kendler, K. S. (2016). The nature of psychiatric disorders. *World Psychiatry, 15*(1), 5–12.

Littlefield, M. M., & Johnson, J. (2012). *The neuroscientific turn: Transdisciplinarity in the age of the brain*. University of Michigan Press.

Logothetis, N. K. (2008). What we can do and what we cannot do with fMRI. *Nature, 453*(7197), 869–878.

Mann, S. L., & Contrada, R. J. (2022). Biological causal beliefs and depression stigma: The moderating effects of first-and second-hand experience with depression. *Journal of Mental Health, 31*(1), 5–13.

Marx, O. M. (1970). Nineteenth-century medical psychology: Theoretical problems in the work of Griesinger, Meynert, and Wernicke. *Isis, 61*(3), 355–370.

Mayes, R., & Horwitz, A. V. (2005). DSM-III and the revolution in the classification of mental illness. *Journal of the History of the Behavioral Sciences, 41*(3), 249–267.

Meynert, T. (1884). *Psychiatrie: Klinik der Erkrankungen des Vorderhirns begründet auf dessen Bau, Leistungen und Ernährung*. Wilhelm Braumüller.

Mirowsky, J., & Ross, C. E. (2003). *Social causes of psychological distress* (2nd ed.). De Gruyter.

Moncrieff, J., Cooper, R. E., Stockmann, T., Amendola, S., Hengartner, M. P., & Horowitz, M. A. (2023). The serotonin theory of depression: A systematic umbrella review of the evidence. *Molecular Psychiatry, 28*(8), 3243–3256.

O'Connor, C., & Joffe, H. (2013). How has neuroscience affected lay understandings of personhood? A review of the evidence. *Public Understanding of Science, 22*(3), 254–268.

OECD [Organization for Economic Cooperation and Development]. (2011). *Sick on the job: Myths and realities about mental health and work*. Organization for Economic Cooperation and Development.

Pickersgill, M., & Van Keulen, I. (Eds.). (2011). *Sociological reflections on the neurosciences*. Emerald.

Pieters, T. (2023). The imperative of regulation: The co-creation of a medical and non-medical US opioid crisis. *Psychoactives, 2*(4), 317–336.

Pormann, P. E. (Ed.). (2008). *Rufus of Ephesus: On melancholy*. Mohr Siebeck.

Racine, E., Dubljević, V., Jox, R. J., Baertschi, B., Christensen, J. F., Farisco, M., et al. (2017). Can neuroscience contribute to practical ethics? A critical review and discussion of the methodological and translational challenges of the neuroscience of ethics. *Bioethics, 31*(5), 328–337.

Racine, E., Waldman, S., Rosenberg, J., & Illes, J. (2010). Contemporary neuroscience in the media. *Social Science & Medicine, 71*(4), 725–733.

Rasmussen, M. L., Haavind, H., & Dieserud, G. (2018). Young men, masculinities, and suicide. *Archives of Suicide Research, 22*(2), 327–343.

Rasmussen, N. (2008). *On speed: The many lives of amphetamine*. NYU Press.

Rose, D., & Rose, N. (2023). Is 'another' psychiatry possible? *Psychological Medicine, 53*(1), 46–54.

Roskies, A. L. (2007). Neuroethics beyond genethics: Despite the overlap between the ethics of neuroscience and genetics, there are important areas where the two diverge. *EMBO Reports, 8*(S1), S52–S56.

Rugulies, R., Aust, B., Greiner, B. A., Arensman, E., Kawakami, N., LaMontagne, A. D., & Madsen, I. E. (2023). Work-related causes of mental health conditions and interventions for their improvement in workplaces. *The Lancet, 402*(10410), 1368–1381.

Schleim, S. (2008). Moral physiology, its limitations and philosophical implications. *Jahrbuch für Wissenschaft und Ethik, 13*, 51–80.

Schleim, S. (2014). Turning our attention to the neuroscience turn. *BioSocieties, 9*(3), 354–359.

Schleim, S. (2015). The half-life of the moral dilemma task: A case study in experimental (neuro-)philosophy. In J. Clausen & N. Levy (Eds.), *Handbook of neuroethics* (pp. 185–199). Springer.

Schleim, S. (2020a). To overcome psychiatric patients' mind-brain dualism, reifying the mind won't help. *Frontiers in Psychiatry, 11*, 605.

Schleim, S. (2020b). Neuroenhancement as instrumental drug use: Putting the debate in a different frame. *Frontiers in Psychiatry, 11*, 567497.

Schleim, S. (2021). Neurorights in history: A contemporary review of José MR Delgado's "Physical Control of the Mind" (1969) and Elliot S. Valenstein's "Brain Control" (1973). *Frontiers in Human Neuroscience, 15*, 703308.

Schleim, S. (2022). Why mental disorders are brain disorders. And why they are not: ADHD and the challenges of heterogeneity and reification. *Frontiers in Psychiatry, 13*, 943049.

Schleim, S. (2023). *Mental health and enhancement: Substance use and its social implications*. Palgrave Macmillan.

Schleim, S., & Roiser, J. P. (2009). FMRI in translation: The challenges facing real-world applications. *Frontiers in Human Neuroscience, 3*, 845.

Schleim, S., Spranger, T. M., Erk, S., & Walter, H. (2011). From moral to legal judgment: The influence of normative context in lawyers and other academics. *Social Cognitive and Affective Neuroscience, 6*(1), 48–57.

Scull, A. (2021). American psychiatry in the new millennium: A critical appraisal. *Psychological Medicine, 51*(16), 2762–2770.

Sheldon, E., Simmonds-Buckley, M., Bone, C., Mascarenhas, T., Chan, N., Wincott, M., et al. (2021). Prevalence and risk factors for mental health problems in university undergraduate students: A systematic review with meta-analysis. *Journal of Affective Disorders, 287*, 282–292.

Shorter, E. (2015). The history of nosology and the rise of the Diagnostic and Statistical Manual of Mental Disorders. *Dialogues in Clinical Neuroscience, 17*, 59–67.

Snelders, S., & Pieters, T. (2011). Speed in the Third Reich: Methamphetamine (Pervitin) use and a drug history from below. *Social History of Medicine, 24*(3), 686–699.

Sperry, R. W. (1981). Changing priorities. *Annual Review of Neuroscience, 4*(1), 1–16.

Swami, V., Stanistreet, D., & Payne, S. (2008). Masculinities and suicide. *The Psychologist, 21*(4), 308–311.

Szasz, T. (1974). *Ceremonial chemistry: The ritual persecution of drugs, addicts, and pushers.* Anchor Press.

Tire, C. V. K. (2001). Introduction to special issue on NeuroLaw. *NeuroRehabilitation, 16*, 67–68.

Valenstein, E. S. (1973). *Brain control: A critical examination of brain stimulation and psychosurgery.* Wiley.

Valenstein, E. S. (1998). *Blaming the brain: The truth about drugs and mental health.* The Free Press.

van der Eijk, P. J. (2005). *Medicine and philosophy in classical antiquity: Doctors and philosophers on nature, soul, health and disease.* Cambridge University Press.

WHO [World Health Organization]. (2004). *Prevention of mental disorders: Effective interventions and policy options.* World Health Organization.

Williams, S. J., Seale, C., Boden, S., Lowe, P., & Steinberg, D. L. (2008). Waking up to sleepiness: Modafinil, the media and the pharmaceuticalisation of everyday/night life. *Sociology of Health & Illness, 30*(6), 839–855.

Wilson, M. (1993). DSM-III and the transformation of American psychiatry: A history. *American Journal of Psychiatry, 150*, 399–399.

Zeki, S., & Goodenough, O. (2004). Law and the brain: Introduction. *Philosophical Transactions of the Royal Society B: Biological Sciences, 359*(1451), 1661–1665.

Zilboorg, G., & Henry, G. W. (1969). *A history of medical psychology.* W. W. Norton.

CHAPTER 2

Psychological and Brain Development

There is no such thing as an average adolescent. (developmental neuro-scientist Sarah-Jayne Blakemore, University College London; quoted in Ledford, 2018, p. 431)

Likewise in youth, because of the process of growth, people are in a state similar to drunk, and youth is pleasant. (Aristotle, 4th century BCE; as translated by van der Eijk, 2005, p. 152)

Traditionally, people have been divided into children and adults according to their age. The former are dependent on their parents or other guardians and are generally less responsible before the law, while the latter have more autonomy but also more responsibility. In addition to the category of puberty, which is primarily related to physical sexual maturity, the development of young people is categorized as adolescence and, more recently, "emerging adulthood" (Arnett, 2000; Ryan, 2019). These distinctions are explained in more detail in the next section.

Age differences are of great importance for the law (e.g. Boni-Saenz, 2022; Ryan, 2019). However, different legal systems draw very different boundaries in some cases. For example, Japan reduced the voting age from 20 to 18 a few years ago, while in Scotland it is 16 (Sawyer et al., 2018). Some academics generally consider 16 to be the appropriate voting age (Nelkin, 2020). As this book was being written, voters in Germany

© The Author(s) 2025
S. Schleim, *Brain Development and the Law*, Palgrave Studies in Law, Neuroscience, and Human Behavior,
https://doi.org/10.1007/978-3-031-72362-9_2

were allowed to vote for the members of the European Parliament from this age for the first time. In the EU, only Austria, Belgium and Malta have this low age limit for European elections, while they sometimes set different limits for their own national parliaments.

The age of criminal responsibility is another relevant example. However, the legal systems provide very different answers as to the age at which someone can be made responsible for criminal offenses. According to the extensive overview by legal scholar Don Cipriani, the possibilities vary widely: For example, in Cuba, Malaysia, Poland and France there is no lower limit and in the latter country it is weighed up on a case-by-case basis; in Egypt, India, Jordan and Singapore it is seven years; in Belgium, Brazil, Canada, Mexico, the Netherlands, Portugal and Turkey it is twelve; in Bulgaria, China, Germany, Italy, Peru and Spain it is 14; and in Argentina, the Czech Republic, Denmark, Finland, Norway and Iceland, for example, it is only 15 or 16 (Cipriani, 2009; see also Mercurio et al., 2020). The many complex exceptions go beyond the scope of this book.

In the first chapter, we looked at the history of psychiatry, neuroethics and neurolaw. Juvenile justice has also developed over time: The criminologist Jean Trépanier referred to the first efforts in Australia in 1889 and in Norway in 1896; however, he named Chicago in 1899 as the location of the first real juvenile court (Trépanier, 2018; see also Steinberg, 2009). He then distinguished the periods from 1900 to 1930 for the general emergence, 1930–1960 for the consolidation and 1960–1990 for the transformation of international juvenile justice.

However, there can be major differences not only between countries, but also within a legal system. In Germany, for example, more than 40 different age limits can be identified in the applicable laws, from the moment of conception (e.g. in inheritance law), through birth to the age of 80 (e.g. regulations on missing persons).[1] We do not need to discuss the meaning of these differences here. But the above examples clearly show that age is legally relevant. The cases discussed in detail in Chapter 4 will further illustrate this. The purpose of this second chapter is to gain a better understanding of psychological and neurobiological development so that it can later be applied to legal matters.

---

[1] https://de.wikipedia.org/wiki/liste_der_altersstufen_im_deutschen_recht.

2.1   Puberty and Adolescence

In this section, we will look at the—as we will see—more biological concept of puberty and the more psychosocially determined concepts of adolescence and emerging adults. The category of adolescence in particular plays a major role in discussions about legal responsibility. Later on, however, we will also deal with critical positions and address the question of the extent to which the biological or psychological characteristics that are important for the aforementioned terms are fixed for more or less all people or are themselves changeable.

The *New Oxford American Dictionary* defines puberty as "the period during which adolescents reach sexual maturity and become capable of reproduction" and adolescence as "the period following the onset of puberty during which a young person develops from a child into an adult." An adult, in turn, is "a person who is fully grown or developed" or—legally speaking—"a person who has reached the age of majority." In more scientific language, Susan M. Sawyer, Professor of Adolescent Health at the University of Melbourne, and colleagues described puberty as follows:

> Puberty consists of a series of distinct but interlinked hormonal cascades that consist of adrenarche (the activation of adrenal stress hormones that starts between 6 and 9 years of age), the growth spurt, and gonadarche (when pituitary gonadotropins trigger gonadal changes). In well nourished populations, the timing of peak height velocity occurs around age 11 years in girls and 13 years in boys. 50% of girls have evidence of thelarche (breast budding) at age 10 years, and menarche (a late phase of pubertal maturation in girls) occurs around 12-13 years of age. (Sawyer et al., 2018, p. 224)

Puberty is therefore primarily related to the sexual maturity of the body. To this day, this is divided into five stages, called Tanner I to V, based on the studies of James M. Tanner (1920–2010), a British pediatrician (Ledford, 2018). The doctor and his colleagues took nude photos of dozens of young people every few months from 1949 to 1971. They all came from a home for neglected children near London and were exempted from school lessons as a reward for their participation. The differentiation of the stages was based on the growth of the genitals, pubic hair and—only for the girls—the breasts.

The Tanner stadiums are criticized today not only because of the questionable collection and distribution of nude photos of minors. Their dubious representativeness is of greater importance for our scientific question. This is because it has now been shown many times that child and adolescent development depends on social factors such as available nutrition, ethnicity and living in rural or urban regions (Worthman & Trang, 2018).

This can be seen particularly clearly in menarche, girls' first menstruation: Between 1840 and 1940 in Norway, for example, this fell from an average of over 17 to under 14 years; today it averages between 12.5 and 13.5 years in many developed countries (ibid.). In the USA, it fell minimally between 1995 and 2017 from an average of 12.1–11.9 years, and there are differences between ethnic groups (Martinez, 2020). The ever earlier sexual maturity that is sometimes dramatized in the media is therefore more likely the result of less deprivation. For this reason alone, the sample for Tanner's photos was not representative, as these children grew up in worse conditions than others.

Similar changes in connection with individual and social prosperity can be shown for increasing body size in many countries (Worthman & Trang, 2018). It is important for us to note that even a biological-physiological measure does not have to be "objective" in the sense that it is independent of social influences. Although puberty can be scientifically measured and differentiated into various stages, its temporal dynamics are variable. In addition, there are major individual differences: for the years studied from 2013 to 2017, for example, 20% of girls had their first menstruation before their 11th birthday, but the last 20% only after their 13th birthday (Martinez, 2020).

Adolescence

In contrast to puberty, which can be defined by the development of physical sexual characteristics, adolescence is more of a social construct. Some researchers also express this literally (e.g. Arnett, 2000; Ledford, 2018; Worthman & Trang, 2018).

A comment on this: As someone who has been teaching philosophy of science in psychology and the social sciences for around 15 years, I have repeatedly encountered the conviction that social constructs are "not real"—or at least less real than natural objects. However, calling something a "social construct" emphasizes its dependence on social norms and institutions (Berger & Luckmann, 1966/1991; Schleim, 2023). For example, the money we have today exists because it is issued by a central bank and we accept it as a medium of exchange. How could it be used to buy countless goods and services if it were not real? For some people, money is even one of the most important and, in this sense, most real things in life.

We have just seen that the dynamics of puberty also depend on social factors. However, people do not have the physical characteristics of sexual development—such as the size of their sexual characteristics—as a result of attribution by a third party. As we will see in a moment, adolescence is much more related to psychosocial characteristics and what is considered "normal" in a society, especially by its doctors and researchers. There is also less agreement about its beginning and end. In my opinion, all of this justifies talking about adolescence as a social construct. This does not deny that it is ultimately about characteristics that the people described in this way, such as their behavior and their way of life, do not have just because of an attribution.

Figure 2.1 shows that adolescence plays a role in English-language books, especially from the beginning of the twentieth century and with increasing tendency. The term was coined around 1900 by G. Stanley Hall (1844–1924), a developmental psychologist (Arnett, 2006; Hall, 1904). Puberty had already been written about for some time and there was no such sharp rise. Talk of the teenager (from the 1950s) and the "adolescent brain" (from the mid-1990s) came later.

According to the 1970 Tanner criteria, sexual maturity was usually completed around the 15th birthday (Ledford, 2018). As we have seen, this development shifted forward under better living conditions. At this age—15 or younger—most people are still living with their parents or other guardians. Roles that are traditionally associated with adulthood, such as a stable partnership, marriage, parenthood and permanent employment, are usually not yet taken on (Sawyer et al., 2018; Twenge & Park, 2019).

According to anthropological studies, many cultures recognize a transitional phase between childhood and adulthood; however, of 41 societies compared, only 41% (for women) and 35% (for men) had a separate word for this (Schlegel & Barry, 1991; Worthman & Trang, 2018). Certain symbols such as hairstyle, clothing, tattoos and participation in certain rituals then play an important role.

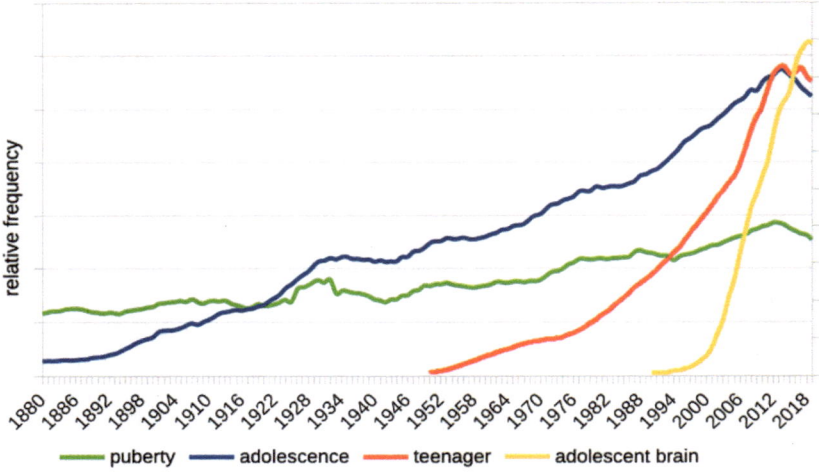

Fig. 2.1 While puberty has been discussed in English-language books for a longer time, the term "adolescence" only became increasingly popular in the twentieth century. The terms "teenager" and "adolescent brain" followed decades later. The first three categories and the last are presented in different scales: "puberty," "adolescence" and "teenager" (10^{-6} percent) occur approx. 50–100 times more frequently than "adolescent brain" (10^{-8} percent) (*Source* Google Ngram)

When Hall described the concept of adolescence as a time of "storm and stress" between the ages of 14 and 24, a period of emotional and behavioral turmoil (Arnett, 2006; Hall, 1904), this is also a testimony to his time and culture. The social change associated with it is the introduction of compulsory education. For example, in the 100 years from 1870 to 1970, the average time spent in schools in the USA increased from under five to over ten years (Lee & Lee, 2016; Worthman & Trang, 2018). As a result, young people enter the labor market later and spend much more time with their peers in the school system.

In the mid-twentieth century, the time from 10 or 11 to 18 or 19 years of age became established in science as the period of adolescence (Arnett, 2000; Sawyer et al., 2018). More recently, however, some have noted that its end point no longer fits with the adoption of the typical adult roles mentioned earlier. For example, the average age at first marriage shifted backwards globally by almost six years (for women) and five years (for men) between 1970 and 2005 (Worthman & Trang, 2018; see also Dahl et al., 2018; Twenge & Park, 2019). Sawyer and colleagues noted that in many European countries, on average, people now marry after their 30th birthday—or live together unmarried more often (Sawyer et al., 2018). The classic pattern of transition from school education to work to marriage and parenthood has been broken.

Different groups of researchers draw different conclusions from these changes (Fig. 2.2). Around the turn of the millennium, developmental psychologist Jeffrey J. Arnett from Clark University in Massachusetts coined the term "emerging adulthood" with resounding success (Arnett, 2000). This refers to the "developmental period between adolescence and young adulthood" at an age "roughly from 18 to 25 years" and "before settling into a career and stable relationship" (Sussman & Arnett, 2014, pp. 147–148). Others, however, proposed a new concept of adolescence that encompasses the entire span from 10 to 24 years (Ledford, 2018; Sawyer et al., 2018).

The shifting age boundaries and different perspectives fit in with the idea of these categories as social constructs: They are not only dependent on the adoption of classic adult roles, but also on the understanding of what is supposed to be "normal" for adults in the first place. Accordingly, Arnett concluded: "Like adolescence, emerging adulthood is a period of the life course that is culturally constructed, not universal and immutable" (Arnett, 2000, p. 470).

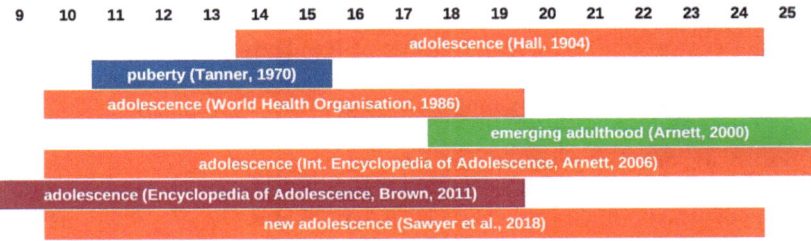

Fig. 2.2 Overview of the age ranges associated with the various categories. The representation is for illustrative purposes; the boundaries and transitions are fluid in individual cases, and the colors serve primarily as a contrast (*Sources* After Arnett [2000, 2006], Brown and Prinstein [2011], Ledford [2018] and Sawyer et al. [2018])

From a scientific point of view, however, this also raises the question of the onset of adulthood, from which the other terms are ultimately differentiated. From a psychiatric perspective: "Assigning an age for the onset of adulthood is a sociolegal construct. Science cannot assign an exact age to adulthood" (Wakefield & McPherson, 2021, p. 163). Accordingly, there is a fixed concept neither of adolescence nor of adulthood. It is important to bear this in mind below when ascribing certain characteristics to such phases of life, including in the actual legal examples in Chapter 4.

We can summarize this by saying that the beginning of adolescence today is usually defined as the beginning of puberty, which has shifted forward, and its end is pragmatically defined as the assumption of typical adult roles, which has shifted backwards. To conclude, here are two definitions from the *Encyclopedia of Adolescence* by B. Bradford Brown and Mitchell J. Prinstein and the *International Encyclopedia of Adolescence* by Jeffrey J. Arnett:

> Adolescence: A stage in human life cycle covering the years after the onset of puberty until the onset of adulthood (approximately ages 9-19 years). The adolescent phase is characterized by a growth spurt in height and weight, the development of secondary sexual characteristics, sociosexual maturation, and intensification of interest and practice in adult social, economic, and sexual activities. (Bogin, 2011, p. 275)

And:

> Scholars view adolescence as beginning with puberty, and age 10 is when the first outward signs of puberty occur for most girls in industrialized countries (boys usually begin about 2 years later). [...] Setting the upper age boundary of adolescence is more difficult and more subject to cultural variability. Scholars generally view adolescence as ending when adulthood begins, which sounds simple enough – until one tries to answer the question of when adulthood begins. [...] Age 25 was chosen as the upper boundary partly for practical reasons. (Arnett, 2006, p. viii)

Psychological Characteristics

What psychological characteristics are usually associated with adolescence? Hall famously coined the image of a phase of "storm and stress." Around

100 years later, scientists spoke, for example, of an increased willingness to take risks, increased substance use and other dangerous behaviors (Crone & Dahl, 2012). More positive views portrayed adolescence or emerging adulthood as a time of finding one's identity and trying things out as well as learning and adapting quickly (Arnett, 2000; Dahl et al., 2018).

The developmental psychologist Laurence Steinberg from Temple University in Philadelphia summarized psychological development in adolescence as follows: Cognitive understanding and reasoning develop between the ages of about eleven and 16. During this period, abstract, deliberative and hypothetical thinking improves until, at around the age of 16—at least under experimental conditions—there are no longer any significant differences to adults (Steinberg, 2009).

However, Steinberg and his colleagues differentiated this from situations in which emotional and social aspects play a greater role: According to them, adolescents are particularly susceptible to peer pressure, immediate versus delayed rewards, are less future-oriented as a result and are less able to control their impulses (Steinberg, 2009; Steinberg et al., 2009; see also Cohen et al., 2016). Deficits in these areas could be identified into young adulthood, particularly for difficult situations.

These researchers have demonstrated this experimentally with a game called "Tower of London." The test subjects have to move a configuration of colored balls to a predetermined target position, but are only allowed to move one ball per round. If mistakes are made, these must be corrected with additional moves. If the goal is easy to achieve, there are no differences between adolescents and adults, unlike with difficult tasks (Steinberg et al., 2008). In other tasks, adolescents performed well if they had to solve a problem alone; in the presence of peers, however, they would take excessive risks, probably to impress others (Steinberg, 2009).

With regard to such deficits, development researcher Ronald E. Dahl from the University of California in Berkeley and colleagues summarized the problems of this phase of life as follows:

> As supported by a large number of studies, this developmental trajectory from childhood to emerging adulthood is fraught with a multitude of risks and vulnerabilities. These contribute to a marked increase in risk of death and disability through adolescent accidents, suicide, violence, depression,

alcohol and substance use, sexually transmitted diseases, unwanted pregnancies, as well as the establishment of a wide range of health-related behavioral risk factors (such as smoking, drinking, substance use, unhealthy eating and sedentary behavior) that will contribute to health consequences in later life. (Dahl et al., 2018, p. 442)

A new study by behavioral and neuroscientists led by Beatriz Luna from the University of Pittsburgh with data from over 10,000 test subjects aged 8–35 years partially confirmed the picture presented, but also contradicted it. In the experiments, the researchers focused on executive functions such as working memory, planning ahead, attention and suppression of premature reactions (Tervo-Clemmens et al., 2023). The evaluation of their own studies and those of other research groups showed a strong improvement in cognitive performance between the ages of 10 and 15, which finally converged with the results of the older test subjects between the ages of 18 and 20 (Fig. 2.3). This applied to both the accuracy of the answers and the reaction speed.

This means that, on average, cognitive development is largely complete from the age of 16 and more or less complete from the age of 18—including impulse control. However, the researchers discussed the limitation that emotional information processing was not examined in detail in the experiments (Tervo-Clemmens et al., 2023). In addition, such studies in the behavioral psychology laboratory always raise the question of transferability to everyday life: There may be greater differences between adolescents and adults in socially complex situations or under peer pressure that are not so easy to demonstrate experimentally, at least not in such a large sample to date. The data analyzed by Tervo-Clemmens and colleagues also came from subjects in the USA, but reflected the country's ethnic diversity. Furthermore, such mean values cannot be transferred to individual cases without restriction.

It is therefore not just a complex question of when adolescence begins and at what age it ends. A conclusive assessment of the psychological similarities and differences with other age groups requires further research. This is made more difficult by the shifting boundaries, because adolescence is not a concrete thing, but changes itself as a result of social and scientific trends.

The latter was also shown by an analysis of data from almost 8.5 million US Americans aged 13–19. According to this, the frequency of typical adult activities—understood here as having a driver's license, trying

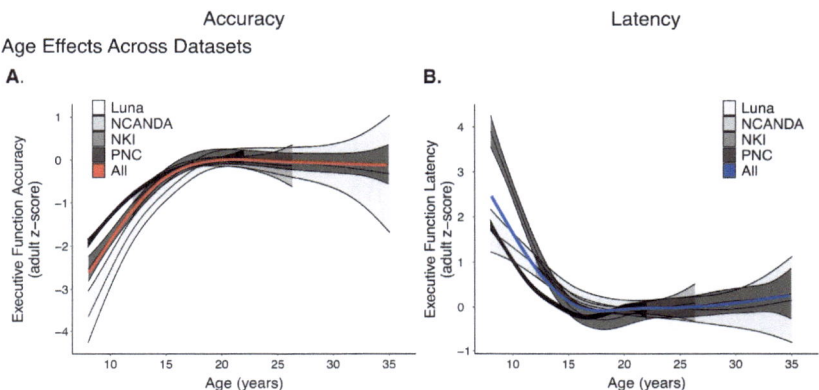

Fig. 2.3 According to the large-scale study by Tervo-Clemmens and colleagues, the development of executive functions can be divided into three age groups: from 10 to 15 years, both accuracy (left, A) and reaction speed (right, B) improve strongly; at 16–17 years, the results approach the final state asymptotically, which is reached at 18–20 years. Luna, NCANDA, NKI and PNC denote four independent data sources. The red and blue lines represent the average of all studies; the gray areas represent the mean and variance of the four data sources (*Source* Tervo-Clemmens et al. [2023]. License: CC BY 4.0 [http://creativecommons. org/licenses/by/4.0/])

alcohol, pursuing paid work, going out without parents and dating—decreased continuously between 1976 and 2016 (Twenge & Park, 2019). The researchers described this as a "slower life strategy" and concluded: "The developmental trajectory of adolescence has slowed, with teens growing up more slowly than they once did" (ibid., p. 653). Or to quote a headline from *Scientific American*: "Extended Adolescence: When 25 Is the New 18."[2]

We are therefore dealing with two opposing trends here: On the one hand, sexual maturity and thus the age of puberty have decreased in recent decades, while on the other hand, typically adult behavior and life patterns have emerged later and later (see also Dahl et al., 2018; Sawyer et al., 2018). Accordingly, the intermediate phase known as "adolescence" or

[2] https://www.scientificamerican.com/article/extended-adolescence-when-25-is-the-new-181/.

"emerging adulthood" is being extended. This will be important for the discussion of legal issues in Chapter 4. In this chapter, however, we will first deal with brain development and, finally, with two critical objections to the concept of adolescence.

2.2   Major Phases of Brain Development

The brain is the center of the human nervous system and consists of the brain stem, cerebellum, diencephalon and cerebrum. Compared to other species, the large and very efficiently folded surface of the latter, the cerebral cortex, is striking in humans.

The number of neurons in the brain has long been estimated at 100 billion, but according to more recent estimates, the average adult has around 86 billion. Of these, however, only around 19% or 16 billion are located in the cerebral cortex and most of the rest in the cerebellum (Herculano-Houzel, 2009). But the structure of the nerve cells in the latter is much less complex. In addition to neurons, there are also other cells in the brain that have a supply function but perhaps also a role in information processing. Glial cells are an example of this. Their number has been estimated to be ten times that of nerve cells, i.e. around one trillion. According to recent investigations, however, there are probably slightly fewer glial cells in the brain than neurons (Von Bartheld et al., 2016).

These widely differing estimates are an example of the fact that we are still far from knowing everything about the so-called most complex object known to us in the universe. Basic research is even continuing into the number of cellularly distinguishable brain regions in the cerebrum. In 2020, 30 years after the start of the "Decade of the Brain," around 70% of the cerebral cortex had been fully mapped and the number of areas was estimated at over 180 (Amunts et al., 2020). By comparison, cognitive neuroscientists still often use a brain map that is more than a hundred years old and identifies only 43 brain regions, the so-called Brodmann areas (Zilles & Amunts, 2010).

This focus on what we do not yet (fully) know about the brain should also make us cautious when making statements about its development. The finding in the previous section that the age limit of adolescence is shifting does not make the situation any easier. As explained at the beginning, this book takes the view that our perception, feeling, thinking and behavior arise from a combination of body, brain and situation. If experts now change their opinion about how early adolescence begins and when it ends, this may also be due to situational and cultural factors. But of course these factors in turn have an effect on the body and brain, just as reading this book is changing your brain right now. In the following, we want to get an understanding of what recent research has to say about the development of the cerebrum.

Gray and White Matter

An important difference is that between gray and white matter. The former is found where the cell nuclei of the neurons occur most frequently, namely in the cerebral cortex, the folded surface of the cerebrum; the latter is the substance in between, through which the nerve cells connect with others over long distances. An important principle is the myelination of these connections: A layer of fatty tissue, which explains the white color, provides better electrical insulation for the nerve fibers. This significantly increases the speed of signal transmission. A fundamental feature of brain development is then, on the one hand, a *decrease* of gray matter with simultaneous specialization of local connections and, on the other hand, an *increase* of white matter due to the expansion of distant connections in the cerebrum (Bigler, 2021; Fair et al., 2009; Somerville, 2016). It is generally consistent with this that the brain grows very quickly at the beginning of our lives and then becomes smaller again over time. We will discuss this in more detail in the next few paragraphs.

At the age of only six years, the brain is already about 90% of its adult size; the thickness of almost all areas of the cerebral cortex reaches its maximum before the age of about 10.5 years (Tamnes et al., 2010). However, there are important differences here and areas related to basic functions such as perception and movement mature faster than those related to social cognition and abstract perception (Sydnor et al., 2021). In a species such as humans, this gradual development fits with the fact that children are initially completely dependent on their parents and only gradually—both literally and proverbially—stand on their own two feet and eventually assume more personal responsibility in social contexts.

Only recently has there been a developmental table for the brain similar to the tables for normal weight and height or the stages of puberty. Richard A. I. Bethlehem from Cambridge University and colleagues were able to compile this table from more than 100 studies using magnetic resonance imaging. In fact, 123,984 MRI images of 101,457 people were analyzed from before birth to old age (Bethlehem et al., 2022). The most important results are summarized in Fig. 2.4.

The figure presents us with an initial challenge, as it shows seven criteria that can be (approximately) measured with MRI: the volume of gray and white matter, subcortical regions, the ventricles, the entire brain, the mean cortical thickness and the total surface area. The ventricles filled with cerebrospinal fluid first increase in size up to the age of two, then remain constant until around the age of 30 and then increase slightly until the sixth decade of life, finally becoming exponentially larger. This gives us no indication of adolescence.

The volume of gray matter peaks at an average age of 5.9 years, the subcortical regions at 14.4 years and the white matter at 28.7 years. Mean cortical thickness even peaks at just 1.7 years, total brain surface area at 11.0 years and total brain volume at 12.5 years (Bethlehem et al., 2022). Like other researchers previously, they found differences between regions, with almost all identified brain areas having reached their maximum volume before the 12th birthday. All of these figures are mean values and the data was predominantly from Western countries.

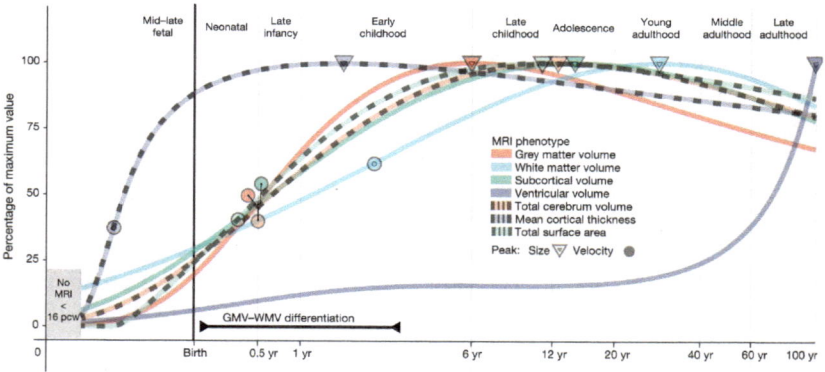

Fig. 2.4 The graph shows a variety of neurobiological changes over the course of a human life, based on 123,984 brain scans, from before birth to old age. Of particular interest to us are the peaks in gray matter volume (red) and white matter volume (light blue), marked by the inverted triangles, at 5.9 and 28.7 years, respectively (*Source* Bethlehem et al. [2022]. License: CC BY 4.0)

If you look at the brain as a whole, you will not find any variable that corresponds to the general understanding of adolescence. After a summarizing statement by Leah Somerville, in the next section we will look in more detail at individual brain regions and the individual differences between people:

> [T]here is little agreement among basic scientists on what properties of a brain should be evaluated when judging whether a brain is mature. This lack of consensus could reflect the fact that most neuroscientists are typically focused on the 'journey' – the temporal unfolding of a particular development process-more than when a brain reaches a particular 'destination.' [...] Some neuroscientists may believe that the very notion of defining brain maturity is a misguided objective, as the brain never stops changing across the entire lifespan. (Somerville, 2016, pp. 1164, 1166)

Diversity and Variability

As already mentioned, not all brain regions develop at the same pace. In particular, those associated with social cognition and abstract thinking generally develop later (Sydnor et al., 2021). The psychosocial deficits that adolescents are sometimes accused of, as we saw in Sect. 2.1, fit in with this. However, even on closer inspection, the connection between brain and behavior is not as clear-cut as one might wish for a neurobiological underpinning of normative age limits.

For example, the psychologist Christian K. Tamnes from the University of Oslo and colleagues calculated another parameter, the index for "fractional anisotropy" (Tamnes et al., 2010). This is related to various neurobiological characteristics, such as the length of the cell connections, the neuronal density and their myelination. By measuring these and other values, they found changes in some regions of the frontal brain even after the age of 30. However, they only examined 168 people between the ages of eight and 30 for their study, so in principle they were unable to draw any conclusions about longer-lasting processes. Another study even found that the process of myelination in humans intensifies again toward the middle of the third decade of life (Miller et al., 2012).

In addition to this abundance or diversity of ways to describe brain development, we must also consider the challenge of individual differences. Every brain is unique, which is why neuroscientists put a lot of effort into developing probability maps (Amunts et al., 2020). Particularly in individual cases, such as brain surgery, we cannot simply assume that a certain coordinate corresponds to the same area in all people. The brains of men also have a larger volume and greater variability on average than those of women (Bethlehem et al., 2022).

Differences in brain development have already been linked to psychosocial variables, such as socioeconomic status (Piccolo et al., 2016). Figure 2.5 gives an impression of this.

More important for us than the interpretation of such group differences is the variability *within* a group. For example, you can look at the lowest scores of some six-year-olds: These fall on the mean of 13- to 14-year-olds. Or look at the highest scores of some 18-year-olds: These fall on the mean of twelve-year-olds (Piccolo et al., 2016).

In another study, psychologists Christopher R. Madan and Elizabeth A. Kensinger from Boston College attempted to solve the problem with artificial intelligence. They used an algorithm to recognize patterns in the cortical structures of 1056 people between the ages of 18 and 97 based on their MRI scans (Madan & Kensinger, 2018). It was then possible to determine a person's age based on their brain image with an average error of six to seven years. The researchers call this result a "reliable age prediction" (ibid., p. 399). However, a few years can make a big difference, especially in young people. This experiment should therefore be repeated with more refined methods and younger test subjects in order to check the accuracy rate for adolescents; this should be feasible based on the data from Bethlehem and colleagues.

However you look at it: It is clear that brains are plastic and constantly changing; although certain peak and turning points can be determined mathematically, they point in different directions. It therefore seems impossible to define the concept of adolescence in neurobiological terms for the time being. In any case, the data discussed here are not compatible with any fixed age limits, as they show fluid transitions at all moments. Furthermore, it is estimated that cortical thickness changes by less than -1.25% per year from childhood onwards and by less than -0.75% per year from the twenties onwards (Amlien et al., 2016).

All of this suggests a fluid quantitative, but not a hard qualitative distinction, as the classification of people as children, adolescents and adults requires—especially in law. In the case of legal age limits, a tidal change can occur within a second, for example on a birthday when a person reaches the age of majority. Biological changes rarely take place

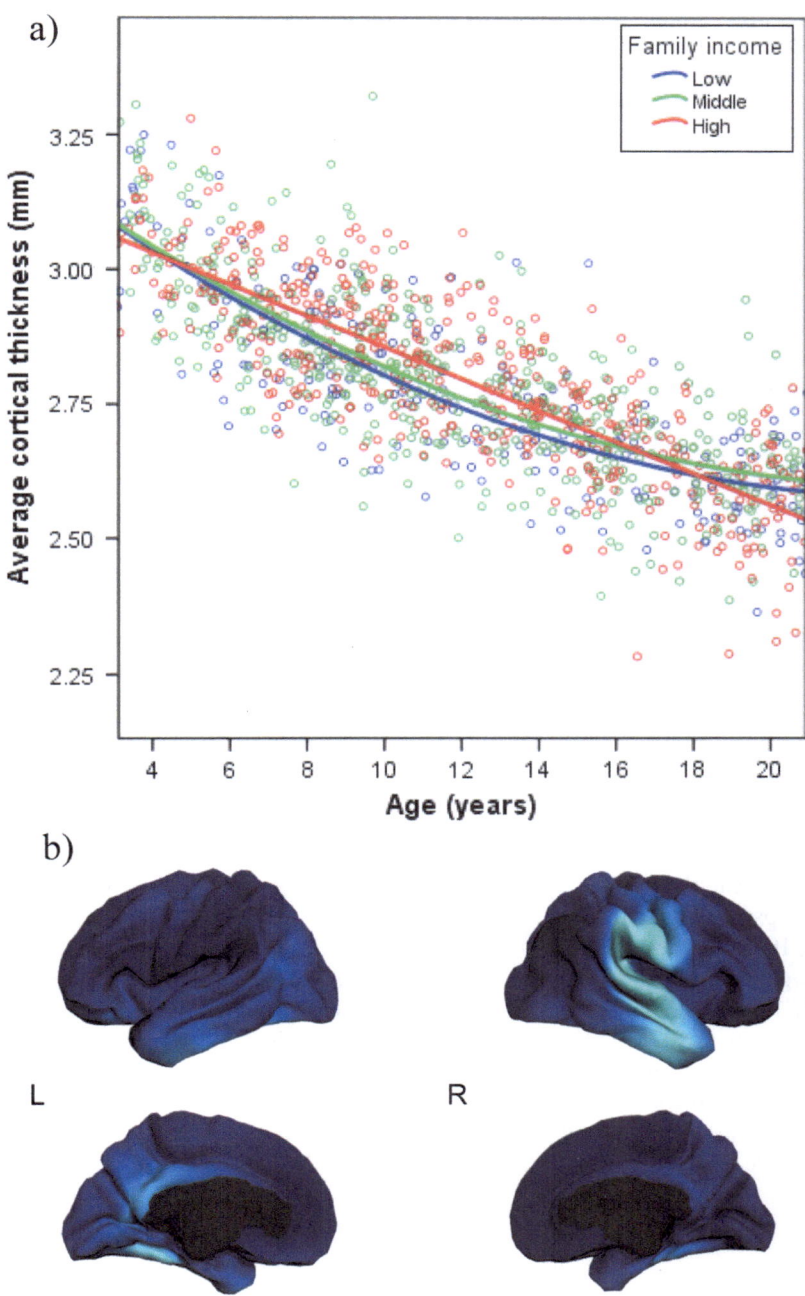

◄**Fig. 2.5**   Piccolo and colleagues related the socioeconomic status of 1148 individuals aged three to 20 years to mean cortical thickness and age. The differences between the three groups from families with an annual income of $4500–$25,000 (blue), $35,000–$75,000 (green) and $125,000–$325,000 (red) were perhaps small (top), but statistically significant at the $p < 0.001$ level for the brain areas highlighted in light blue below (*Source* Piccolo et al. [2016]. License: CC BY 4.0 [http://creativecommons.org/licenses/by/4.0/])

in this way. With the publication by Bethlehem and colleagues, however, there is now at least a benchmark that can be used for brain development. With this important observation in mind, we conclude this chapter by looking at how the concept of adolescence can be criticized from a psychosocial perspective.

2.3   AGAINST ADOLESCENCE

In Sect. 2.1, I argued—similarly to other researchers—that the category of "adolescence" is a social construct, because the distinction depends to a large extent on what certain actors understand by a "normal" adult life. Accordingly, leading researchers in this field are adjusting the age boundaries because typical behavioral patterns are changing. However, this does not say anything about the extent to which this behavior itself reflects a natural development process—or is perhaps rather a reaction to social circumstances.

We have already briefly explored the idea that the understanding of adolescence is linked to the introduction of compulsory education. In this section, we now delve deeper into the hypothesis that the specific behavior of young people is primarily a reaction to how (especially: Western) societies treat them. This type of argument has been strongly advocated by the psychologist Robert Epstein.

If G. Stanley Halls *Adolescence* of 1904 is a kind of manifesto for this stage of life, then Epstein's *The Case Against Adolescence: Rediscovering the Adult in Every Teen* could be seen as the corresponding anti-manifesto (Epstein, 2007a). His strategy is threefold: He shows that, firstly, the typical behavior of adolescents today has developed historically; secondly, that even in today's (Western) societies, adolescents can behave like adults if they are allowed to; and thirdly, that young people of the same age behave differently in other societies.

We have already seen that the term "adolescence" began to appear in English-language books around 1900 and then gained in importance over the course of the twentieth century (Fig. 2.1). Epstein argued from a historical perspective that the meaning of "child" did not originally refer to an age group, but to a kinship relationship (Epstein, 2007a). For a long time, children were regarded as a kind of small adult and worked together with adults as early as possible. Over time, the legal limit for the protection of children and young people was raised from under eight years of age, as in Great Britain in the eighteenth century, to 10, 14, 16 and in some cases even up to 18 or 21 years of age.

Before the twelfth century, there were hardly any depictions of children in art as we know them today, but at best of small people in adult clothing (Ariès, 1962; but see Orme, 2001). The restriction or outright prohibition of child and adolescent labor—partly driven by labor unions during poor economic times to reduce competition from cheaper child and youth labor—only gradually allowed a "teenage culture" to emerge (Epstein, 2007a).

Epstein counted the number of laws in the USA that specifically restrict the behavior of people under the age of 18. Before 1800, there were virtually no such laws, by the turn of the century around 1900 there were less than two dozen, and since the 1960s in particular, these have risen dramatically to over 140 (Epstein, 2007a, 2007b). In Chapter 4, we will see that that number has even become much bigger. According to Epstein, many of such laws were more moral in nature and concerned leisure activities such as going to the movies or dancing or staying out after a certain time. In his words:

> Note that virtually all of these newly defined crimes had no victims. They were, if anything, 'crimes against oneself' – at least from the perspective of the authorities. What's more, most of these indiscretions were typical of the working class and poor, which raises questions about the motives of the some of the leaders who fought for such laws. (Epstein, 2007a, p. 44)

In this context, he also spoke of the infantilization of young people. As counter-examples, he cited people from our contemporary history who took on positions of responsibility at the age of 18 or 19, for example as mayors (Epstein, 2007b). Such examples also existed in the past. In pre-industrial societies, however, there was usually neither a word for adolescence nor the problematic behavior typical of it. In our time, too, it is noticeable that this behavior is less pronounced in other cultures, but is increasing to the extent that a Western lifestyle is gaining acceptance there—for example through media use (Epstein, 2007a, 2007b). It is therefore not surprising that Epstein considered the term "teenage brain" to be particularly problematic.

We keep his criticism in mind as an alternative hypothesis. In Sect. 2.1, we have already established that the boundary between "typical" adolescent and adult behavior shifts at a speed that makes a biological-genetic explanation seem unlikely; genetic selection does not manifest itself at the pace of decades. In addition, we have already referred to behavioral research, according to which young people make more risk-taking decisions in the presence of peers than when they are alone.

Then it is not entirely without paradox to first force young people into groups with only their peers, as in school, and then criticize them for problematic group behavior. From this perspective, Epstein's thrust *Against Adolescence* has a certain plausibility, even if there are doubts about some of its evidence, such as the depiction of children in art history (Orme, 2001). At the end of this section and chapter, I would like to briefly refer to another example from clinical, developmental or family psychology.

Parentification

From the discussion so far, we can deduce the hypothesis that behavior—including that of adolescents—depends on the psychosocial situation. However, long-term social trends are difficult to capture experimentally, and for obvious ethical reasons, we cannot simply place adolescents in a different cultural context to study its effects on their thinking and behavior.

In brain research, lesion studies in which specific areas of the brain are deliberately destroyed are all the more forbidden. If such damage occurs, for example, due to a hemorrhage, illness or injury, this is sometimes referred to as a "natural experiment." However, the extent of the damage is then more uncontrolled than in lesion studies in animal experiments. This makes the psychological effects more difficult to interpret (Moll & de Oliveira-Souza, 2007).

One could say that "natural experiments" also occur in the psychosocial development of some adolescents. By this I mean the phenomenon known as parentification, especially in certain psychotherapeutic schools, which has been researched for decades (e.g. Chase, 1999; Jurkovic, 1997). While it is normal in families to a certain extent for adolescents to gradually take on more responsibility in and for the family, the tasks must be appropriate to their capabilities. Otherwise, this can lead to overburdening, which can have negative—but perhaps also positive—consequences. By calling this a "natural experiment," one does not automatically endorse it.

A divorce, major problems, serious illness or even the death of a family member can lead to a role reversal. Children then sometimes have to take on tasks that actually belong to their parents. Sometimes such patterns are passed down through the generations. More formally defined: "Parentification is a type of role reversal, boundary distortion, and inverted hierarchy between parents and other family members in which adolescents assume developmentally inappropriate levels of responsibility in the family of origin" (Hooper, 2018, p. 2697).

In research, a distinction is often made between *instrumental* and *emotional* forms of role-taking: The former involves, for example, cooking meals, taking on other household tasks or financial responsibilities; the latter refers to having to support or care for parents or siblings in the emotional sphere or having to resolve family conflicts. It is assumed that the risk of consequential damage such as school and relationship problems or mental disorders is higher with emotional parentification; these consequences can also only occur years later (Hooper, 2018). Neuroscientists largely agree that the adolescent brain is particularly susceptible to negative stimuli such as severe stress (Eiland & Romeo, 2013; Fuhrmann et al., 2015). Accordingly, Ursula A. Tooley and colleagues hypothesized "that greater exposure to chronic stress accelerates brain maturation" (Tooley et al., 2021, p. 372).

However, family therapists are not only critical of the phenomenon, even if for a long time the negative effects have been researched more than possible positive consequences (Hooper et al., 2008). On the one hand, it can stabilize the family as a whole, which may prevent worse consequences; and on the other hand, individuals can also grow from the tasks and responsibilities, develop more resilience and better problem-solving skills. But if a child or adolescent only *appears* to be able to cope with adult tasks when in reality they are not up to them, developmental researchers speak of "pseudomaturity" or "adultoids" (Galambos & Tilton-Weaver, 2000; Hooper, 2018).

Whether the consequences of parentification are predominantly positive or negative in individual cases depends on many factors. As is so often the case, more research is needed. However, the fact that the phenomenon exists is of great interest for our purposes in this chapter: It shows that, under certain circumstances, young people *can* take on roles that are typical of adults. Their behavior is therefore not only determined by the so-called adolescent brain, but also by important environmental factors in the family and society.

This demonstrates once again that not only individual development, but also adolescence is something plastic: At least to a certain extent, young people react to the demands of the situations they find themselves in. In times of crisis and war, this can even affect an entire generation. This also fits with the understanding of adolescence as a social construct.

One could say with the Canadian philosopher of science Ian Hacking (1936–2023) that adolescence is both a *moving target* and an *interactive kind* (Hacking, 1999). These concepts illustrate Hacking's philosophical stance called "dynamic nominalism" (Hacking, 2007). This describes the dynamic relationship between a name (definition, classification) and that which is named (defined, classified). Hacking and others illustrated this with many examples from the life and social sciences. And, as the discussion in this chapter has shown, adolescence is a clear case, too: Its criteria and age limits change. In this sense, adolescence "moves" and is not a fixed thing like silver which is defined as the element with 47 protons—and always will be.

Subsequently, social and institutional practices of adolescence change, not the least in law as we will analyze in more detail in Chapter 4, and with them the behavior of the people classified as adolescents; this, in turn, influences the classification and institutional practices. Besides speaking of "interactive kinds," Hacking called this the "looping effect" (ibid.). Remember the targeted interventions many researchers demand for this group (e.g. Dahl et al., 2018; Somerville, 2016; Tooley et al., 2021; Worthman & Trang, 2018; see also Rose, 2010). Importantly, this does *not* make the behaviors and people classified as such any less real.

With these important conclusions, we can now close the chapter on psychological development and that of the brain.

2.4   Summary

In this second chapter of the book, we looked at the complexity of biological and sociological categories such as puberty and adolescence. As we saw, the onset of the former is shifting forward—probably due to better life circumstances—and the boundary of the latter is shifting backward—due to changes in young people's lifestyles. Pragmatically, it is now proposed to extend adolescence to the age of 10–24, although there are also alternative accounts. In the words of Sawyers and colleagues:

> Age definitions are always arbitrary, and chronological approaches to the definition of adolescence will continue to be shaped by culture and context. However, puberty marks a major point of discontinuity, with the next phase of growth and neurocognitive maturation continuing past 20 years of age. Tied to the widely spread postponement of role transitions to adulthood, our current definition of adolescence is overly restricted. The ages of 10-24 years are a better fit with the development of adolescents nowadays. (Sawyer et al., 2018, p. 5)

There have probably always been stages of sexual maturity for us humans, whether they were called "puberty" or not. After all, we are born immature and our sexual characteristics only fully develop over time. In many countries today, this begins around the age of ten to twelve, slightly earlier for girls than for boys.

Does adolescence exist in the same way? As we briefly touched on in this chapter, not all cultures have a separate word for the phase between childhood and adulthood. In some, this kind of maturation is indicated by external features such as clothing, hairstyle or tattoos—and of course there are also fashions in countries with a term like "adolescence." However, the idea of adolescence depends crucially on what is regarded as typically childlike on the one hand and typically adult on the other. In the course of the twentieth century, many Western cultures established their own term for this, which in turn was taken up by experts from psychology and other social sciences, medicine and now also the neurosciences. The next chapters will look in more detail at the interaction of law with such concepts, particularly with regard to neurobiological data.

Yet, how realistic is the notion of an "adolescent brain" or "teenage brain," if adolescence is a social construct—constructed by social institutions and their professionals to achieve particular ends? In many cases, adolescents may exhibit typical adolescent behavior, but according to the account presented here and especially the sections on *Against Adolescence* and parentification, this behavior is often a reaction to the situations young people find themselves in. In the form of schools, it is even about environments into which young people are now *forced* into in many, if not virtually all countries. In this sense, the problematic behavior of adolescents in terms of "storm and stress" would itself be a social construct.

It does not seem plausible that neurobiology reduces this complexity. In Chapter 1, we discussed the failed attempts to place the hundreds of mental disorders in the American diagnostic manual DSM, for example, which is also widely used in research and other countries, on a neuroscientific foundation. In this chapter, we learned about a variety of characteristics—such as the surface area, thickness, volume of brain areas or more abstract constructs—in which development can be traced over the course of a person's life. Accordingly, important stages of development are already reached before the common legal age of majority at 18 and, in particular, the now proposed end of adolescence. And if minor changes to the nervous system are taken into account, development is never fully complete until death. The plasticity of the brain is an important characteristic for a lifetime of learning and adapting to different requirements, even if this may become more difficult with age.

From the point of view of scientific theory, it is particularly important not to reify the changing concept of adolescence or the teenager on the basis of neurobiological data. As we have seen, these conceptual distinctions depend to a large extent on social norms. With the term "adolescent brain" or "teenager brain," these norms are transferred to the biological realm and then suddenly appear as *natural* categories. This creates the risk of a biologistic or even naturalistic fallacy: From what has been neurobiologically selected and constructed in this sense, conclusions are then drawn about what behavior is *socially* acceptable as supposedly dictated by the "adolescent brain." The brain itself is a plastic object that is continuously shaped by psychosocial conditions.

In all of this, we have not even addressed the issue of individual differences. We saw that for the comparatively easy-to-detect first menstruation, there is a difference of over two years between the earliest and latest 20% of girls (Martinez, 2020). That's a big difference at the age of puberty. The relevant research in the field of brain development has not yet been completed. But we have already seen evidence of even greater variability here. Clear dividing lines are useful for research and also for the law, such as the distinction between children, adolescents and adults or between minors and adults. Laws must be kept general in the interests of the rule of law. It is the daily business of courts to take into account the circumstances of the individual case. We will now discuss some examples of this in detail in the next two chapters.

References

Amlien, I. K., Fjell, A. M., Tamnes, C. K., Grydeland, H., Krogsrud, S. K., Chaplin, T. A., et al. (2016). Organizing principles of human cortical development—Thickness and area from 4 to 30 years: Insights from comparative primate neuroanatomy. *Cerebral Cortex, 26*(1), 257–267.

Amunts, K., Mohlberg, H., Bludau, S., & Zilles, K. (2020). Julich-Brain: A 3D probabilistic atlas of the human brain's cytoarchitecture. *Science, 369*(6506), 988–992.

Ariès, P. (1962). *Centuries of childhood: A social history of family life.* Vintage Books.

Arnett, J. J. (2000). Emerging adulthood: A theory of development from the late teens through the twenties. *American Psychologist, 55*(5), 469.

Arnett, J. J. (2006). Introduction. In J. J. Arnett (Ed.), *International encyclopedia of adolescence* (pp. vii–ix). Routledge.

Berger, P. L., & Luckmann, T. (1966/1991). *The social construction of reality: A treatise in the sociology of knowledge.* Penguin Books.

Bethlehem, R. A., Seidlitz, J., White, S. R., Vogel, J. W., Anderson, K. M., Adamson, C., et al. (2022). Brain charts for the human lifespan. *Nature, 604*(7906), 525–533.

Bigler, E. D. (2021). Charting brain development in graphs, diagrams, and figures from childhood, adolescence, to early adulthood: Neuroimaging implications for neuropsychology. *Journal of Pediatric Neuropsychology, 7*(1), 27–54.

Bogin, B. (2011). Puberty and adolescence: An evolutionary perspective. In B. B. Brown & M. J. Prinstein (Eds.), *Encyclopedia of adolescence* (Vol. 1, pp. 275–286). Elsevier.

Boni-Saenz, A. A. (2022). Legal age. *Boston College Law Review, 63*, 521–569.

Brown, B. B., & Prinstein, M. J. (Eds.). (2011). *Encyclopedia of adolescence* (Vol. 1). Elsevier.

Chase, N. D. (Ed.). (1999). *Burdened children: Theory, research, and treatment of parentification.* Sage.

Cipriani, D. (2009). *Children's rights and the minimum age of criminal responsibility: A global perspective.* Routledge.

Cohen, A. O., Breiner, K., Steinberg, L., Bonnie, R. J., Scott, E. S., Taylor-Thompson, K., et al. (2016). When is an adolescent an adult? Assessing cognitive control in emotional and nonemotional contexts. *Psychological Science, 27*(4), 549–562.

Crone, E. A., & Dahl, R. E. (2012). Understanding adolescence as a period of social–affective engagement and goal flexibility. *Nature Reviews Neuroscience, 13*(9), 636–650.

Dahl, R. E., Allen, N. B., Wilbrecht, L., & Suleiman, A. B. (2018). Importance of investing in adolescence from a developmental science perspective. *Nature, 554*(7693), 441–450.

Eiland, L., & Romeo, R. D. (2013). Stress and the developing adolescent brain. *Neuroscience, 249*, 162–171.

Epstein, R. (2007a). *The case against adolescence: Rediscovering the adult in every teen*. Quill Driver Books.

Epstein, R. (2007b). The myth of the teen brain. *Scientific American Mind, 18*(2), 56–63.

Fair, D. A., Cohen, A. L., Power, J. D., Dosenbach, N. U., Church, J. A., Miezin, F. M., et al. (2009). Functional brain networks develop from a "local to distributed" organization. *PLoS Computational Biology, 5*(5), e1000381.

Fuhrmann, D., Knoll, L. J., & Blakemore, S. J. (2015). Adolescence as a sensitive period of brain development. *Trends in Cognitive Sciences, 19*(10), 558–566.

Galambos, N. L., & Tilton-Weaver, L. C. (2000). Adolescents' psychosocial maturity, problem behavior, and subjective age: In search of the adultoid. *Applied Developmental Science, 4*, 178–192.

Hacking, I. (1999). *The social construction of what?* Harvard University Press.

Hacking, I. (2007). Kinds of people: Moving targets. *Proceedings of the British Academy, 151*, 285–318.

Hall, G. S. (1904). *Adolescence: Its psychology and its relations to physiology, anthropology, sociology, sex, crime, religion, and education* (Vols. I & II). D. Appleton & Co.

Herculano-Houzel, S. (2009). The human brain in numbers: A linearly scaled-up primate brain. *Frontiers in Human Neuroscience, 3*, 857.

Hooper, L. M. (2018). Parentification. In R. J. R. Levesque (Ed.), *Encyclopedia of adolescence* (2nd ed., pp. 2696–2705). Springer.

Hooper, L. M., Marotta, S. A., & Lanthier, R. P. (2008). Predictors of growth and distress following parentification among college students. *Journal of Child and Family Studies, 17*, 693–705.

Jurkovic, G. J. (1997). *Lost childhoods: The plight of the parentified child*. Routledge.

Ledford, H. (2018). Who exactly counts as an adolescent? *Nature, 554*(7690), 429–432.

Lee, J. W., & Lee, H. (2016). Human capital in the long run. *Journal of Development Economics, 122*, 147–169.

Madan, C. R., & Kensinger, E. A. (2018). Predicting age from cortical structure across the lifespan. *European Journal of Neuroscience, 47*(5), 399–416.

Martinez, G. M. (2020). Trends and patterns in menarche in the United States: 1995 through 2013–2017. *National Health Statistics Reports, 146*, 1–11.

Mercurio, E., García-López, E., Morales-Quintero, L. A., Llamas, N. E., Marinaro, J. Á., & Muñoz, J. M. (2020). Adolescent brain development and progressive legal responsibility in the Latin American context. *Frontiers in Psychology, 11*, 522298.

Moll, J., & de Oliveira-Souza, R. (2007). Moral judgments, emotions and the utilitarian brain. *Trends in Cognitive Sciences, 11*(8), 319–321.

Miller, D. J., Duka, T., Stimpson, C. D., Schapiro, S. J., Baze, W. B., McArthur, M. J., et al. (2012). Prolonged myelination in human neocortical evolution. *Proceedings of the National Academy of Sciences, 109*(41), 16480–16485.

Nelkin, D. K. (2020). What should the voting age be? *Journal of Practical Ethics, 8*(2), 1–29.

Orme, N. (2001). *Medieval children*. Yale University Press.

Piccolo, L. R., Merz, E. C., He, X., Sowell, E. R., Noble, K. G., & Pediatric Imaging, Neurocognition, Genetics Study. (2016). Age-related differences in cortical thickness vary by socioeconomic status. *PloS One, 11*(9), e0162511.

Rose, N. (2010). 'Screen and intervene': Governing risky brains. *History of the Human Sciences, 23*(1), 79–105.

Ryan, C. (2019). The law of emerging adults. *Washington University Law Revue, 97*, 1131–1178.

Sawyer, S. M., Azzopardi, P. S., Wickremarathne, D., & Patton, G. C. (2018). The age of adolescence. *The Lancet Child & Adolescent Health, 2*(3), 223–228.

Schlegel, A., & Barry, H. (1991). *Adolescence: An anthropological inquiry*. Free Press.

Schleim, S. (2023). *Mental health and enhancement: Substance use and its social implications*. Palgrave Macmillan.

Somerville, L. H. (2016). Searching for signatures of brain maturity: What are we searching for? *Neuron, 92*(6), 1164–1167.

Steinberg, L. (2009). Adolescent development and juvenile justice. *Annual Review of Clinical Psychology, 5*, 459–485.

Steinberg, L., Albert, D., Cauffman, E., Banich, M., Graham, S., & Woolard, J. (2008). Age differences in sensation seeking and impulsivity as indexed by behavior and self-report: Evidence for a dual systems model. *Developmental Psychology, 44*(6), 1764–1778.

Steinberg, L., Cauffman, E., Woolard, J., Graham, S., & Banich, M. (2009). Are adolescents less mature than adults?: Minors' access to abortion, the juvenile death penalty, and the alleged APA "flip-flop." *American Psychologist, 64*(7), 583–594.

Sussman, S., & Arnett, J. J. (2014). Emerging adulthood: Developmental period facilitative of the addictions. *Evaluation & the Health Professions, 37*(2), 147–155.

Sydnor, V. J., Larsen, B., Bassett, D. S., Alexander-Bloch, A., Fair, D. A., Liston, C., et al. (2021). Neurodevelopment of the association cortices: Patterns, mechanisms, and implications for psychopathology. *Neuron, 109*(18), 2820–2846.

Tamnes, C. K., Østby, Y., Fjell, A. M., Westlye, L. T., Due-Tønnessen, P., & Walhovd, K. B. (2010). Brain maturation in adolescence and young adulthood: Regional age-related changes in cortical thickness and white matter volume and microstructure. *Cerebral Cortex, 20*(3), 534–548.

Tervo-Clemmens, B., Calabro, F. J., Parr, A. C., Fedor, J., Foran, W., & Luna, B. (2023). A canonical trajectory of executive function maturation from adolescence to adulthood. *Nature Communications, 14*(1), 6922.

Tooley, U. A., Bassett, D. S., & Mackey, A. P. (2021). Environmental influences on the pace of brain development. *Nature Reviews Neuroscience, 22*(6), 372–384.

Trépanier, J. (2018). The roots and development of juvenile justice: An international overview. In J. Trépanier & X. Rousseaux (Eds.), *Youth and justice in Western States, 1815–1950: From punishment to welfare* (pp. 17–69). Palgrave Macmillan.

Twenge, J. M., & Park, H. (2019). The decline in adult activities among US adolescents, 1976–2016. *Child Development, 90*(2), 638–654.

van der Eijk, P. J. (2005). *Medicine and philosophy in classical antiquity: Doctors and philosophers on nature, soul, health and disease.* Cambridge University Press.

Von Bartheld, C. S., Bahney, J., & Herculano-Houzel, S. (2016). The search for true numbers of neurons and glial cells in the human brain: A review of 150 years of cell counting. *Journal of Comparative Neurology, 524*(18), 3865–3895.

Wakefield, S. M., & McPherson, P. (2021). How the evolving state of neuroscience informs the definition of adulthood: A psychiatrist's perspective. *Journal of Pediatric Neuropsychology, 7*(4), 161–168.

Worthman, C. M., & Trang, K. (2018). Dynamics of body time, social time and life history at adolescence. *Nature, 554*(7693), 451–457.

Zilles, K., & Amunts, K. (2010). Centenary of Brodmann's map—Conception and fate. *Nature Reviews Neuroscience, 11*(2), 139–145.

CHAPTER 3

The Brain and the Law

Are you going to believe the electroencephalogram, or are you going to believe what you actually saw with your eyes? (an expert witness in Betz v. Travelers Ins., 1955; quoted from Shen, 2016, p. 679)

Up to this point, we have dealt with the historical context and basics of psychology and brain research, with regard to both mental disorders (Chapter 1) and brain development (Chapter 2). In this and the next chapter, we will take a closer look at the legal applications and case studies. In particular, we will focus on the possible interaction of brains and norms in criminal law, for which the discussion is most advanced. However, similar arguments can also be applied to medical law, contract law and electoral law, for example. I will return to this in the outlook in Chapter 5.

After a brief summary of the topics of neurolaw as a whole, we enter into a deeper analysis with both empirical and theoretical examples of the determination of our behavior. This chapter lays important foundations for the interplay between norms and science. We will draw on these in the specific examples from the USA, the Netherlands and Germany in Chapter 4. However, if you are not interested in the thematic breadth and brief history of neurolaw, you can skip the following paragraphs and continue reading at Sect. 3.1 on free will and responsibility.

© The Author(s) 2025
S. Schleim, *Brain Development and the Law*, Palgrave Studies in Law, Neuroscience, and Human Behavior,
https://doi.org/10.1007/978-3-031-72362-9_3

As already mentioned in the introduction, there was a first special issue of a journal on neurolaw in 2001, namely in the then ten-year-old *NeuroRehabilitation*. From today's perspective, however, the thematic focus seems very limited, as it was primarily concerned with the cooperation between doctors and lawyers in court in cases of spinal and brain injuries. The lawyer J. Sherrod Taylor defined neurolaw as "the area of medical jurisprudence concerned with the medical and legal aspects of traumatic brain injury and spinal cord injury" (Taylor, 2001, p. 69). This author is also credited with being the first to use the term "neurolaw." He and colleagues wrote ten years earlier:

> A new type of lawyer is responding to the challenges presented by the growing recognition of neuropsychology in legal circles. The neurolawyer is one who, through interest, education, and training, has developed special expertise in representing clients with traumatic brain injury. (Taylor et al., 1991, p. 294)

In the same year, the lawyer had started publishing a paid newsletter, *The Neurolaw Letter*, which, according to his own information, had over 600 subscribers at its peak (Shen, 2016). The special edition of *NeuroRehabilitation* also dealt with more general issues such as the admissibility of neuropsychological expertise in court proceedings (Stern, 2001).

The aim here is not to write a detailed history of neurolaw (see Shen, 2016). But we have already seen in Chapter 1 a criminological example from the 1930s in connection with psychosurgery, which could be understood as "early neurolaw." In the decades that followed, questions also arose, for example, about the admissibility of electroencephalography (EEG) ·or neuroimaging in court proceedings and about the connection between brain injuries and violent crime. In Sects. 3.1 and 3.2, we will briefly discuss other cases from the 19th to 21th centuries. This introductory section will provide an overview of what is considered to be neurolaw today.

Three years after the special issue in *NeuroRehabilitation* followed one in the *Philosophical Transactions of the Royal Society B: Biological Sciences*. Its articles dealt, among others, with the general question of how neuroscience could change the law (Chorvat & McCabe, 2004; Greene & Cohen, 2004; Jones, 2004), or specifically with the significance for the concepts of responsibility and punishment (Goodenough, 2004). The topic of lie detection using neuroscientific methods has also now been addressed (Spence et al., 2004).

Shortly afterward, we published the German anthology *From Neuroethics to Neurolaw* with a philosophical, neuroscientific and legal focus (Schleim et al., 2007b). The book also dealt with issues such as human nature, psychological and neuroscientific expert opinions in court, lie detection and the (alleged) conflict between determinism, culpability and criminal guilt. Our legal cooperation partner later published the comparative law anthology *International Neurolaw* in which perspectives from various countries were represented (Spranger, 2012).

In a special issue of *Behavioral Sciences and the Law* in 2009, the neurobiological foundations of our decisions were central (e.g. Erickson & Felthous, 2009), in connection with concepts such as empathy and control (Kröber, 2009; Shirtcliff et al., 2009). Particularly noteworthy are an article on the significance of brain lesions for criminal responsibility (Batts, 2009), which we discuss in more detail in Sect. 3.2, and a study on the influence of emotions on jurors' verdicts (Salerno & Bottoms, 2009).

A special issue of the *International Journal of Law and Psychiatry* in 2009 dealt with the neuroscience of aggression, but did not specifically address neurolaw. Three years later, the same journal published an issue that was also dedicated to critical perspectives—also with my participation and using the term "neuroskepticism" (Rachul & Zarzeczny, 2012; Schleim, 2012). Other articles dealt with aspects of brain stimulation and the question of how to deal with incidental findings in neuroscientific research (Heinrichs, 2012; Schmitz-Luhn et al., 2012; Zarzeczny & Caulfield, 2012). A special issue of the *Journal of Criminal Justice* in 2019 dealt specifically with the importance of neurolaw for minors (e.g. Cornet et al., 2019), a topic to be discussed in more detail in the next chapter.

In the now 25 years or so of neurolaw in the narrower sense, there are of course many more monographs, book chapters, edited volumes and journal articles. In Fig. 1.5, we saw the sharp increase since 2000. Some of these publications focused on neuroscience and autonomy, privacy and individual rights (e.g. Blitz, 2017; Blitz & Bublitz, 2021); some on applied forensic criminological issues, such as the individual assessment of culpability or assessment of dangerousness (e.g. Caruso, 2024; Ligthart et al., 2021; Swaab & Meynen, 2023); and some deal with fundamental questions such as the relationship between mind, brain and behavior, determinism and free will, and criminal responsibility (e.g. Hirstein et al.,

Table 3.1 Systematization of neurolaw according to Chandler (2018)

Neurolaw: main category	Subcategory
Law of neuroscience	Legal assessment of interventions in the brain
	Legal assessment of brain injuries
	Legal assessment of the collection and use of brain data
Neuroscience of law	Impact of neuroscience on legal concepts and categories
	Impact of neuroscience on the principles and practices of justice
	Understanding and improving the decision-making of actors in the legal system
	Identification and prediction of legally relevant mental states and behaviors
Self-reflective questions and critical studies	Self-reflective questions
	Critical neurolaw

2018; Pardo & Patterson, 2013, Patterson & Pardo, 2016; Vincent, 2015). Recently, a textbook of almost a thousand pages dealing with all these issues has already appeared in its second edition (Jones et al., 2022).

The aim here is not to be exhaustive, but to give an impression of what neural law is all about. For guidance, it is worth recalling the reviews by Jennifer A. Chandler and colleagues from the University of Ottawa (Chandler, 2018; Chandler et al., 2019). She distinguished, first, legal issues of neuroscience related to brain interventions, brain injury and data collection; second, neuroscientific issues of law, such as the influence of research on legal categories and practices; and, third, self-reflexive questions about the self-understanding of professors and practitioners of law (Table 3.1).

After this brief overview of what neurolaw can be, we will delve deeper into the substantive discussion in the following section. Section 3.1 deals with the question of who or what determines our behavior and how the category of responsibility relates to this. This knowledge is taken up in Sect. 3.2, where we discuss neurological and legal aspects of known cases of brain injury. After a summary in Sect. 3.3, we then continue with normative questions about brain development in Chapter 4.

3.1   FREE WILL, CAUSATION AND RESPONSIBILITY

The question of how we explain our behavior runs through the history of Western philosophy like no other. It was discussed already in Plato's (428/427–348/347 BCE) dialog *Phaedo*. In it, Socrates (469–399 BCE) spoke to his students for the last time shortly before his death—he had to either go into exile or drink from the cup of hemlock due to alleged blasphemy and incitement of youth. There he discussed the views of his former teacher Anaxagoras (c. 499–428 BCE), a materialist natural philosopher: How much sense does it make to answer the question of why Socrates was in prison with: "because the tendons and bones moved in such-and-such a way"? From today's perspective, we would call this "reductionism" and contrast it with "acting for reasons" (Dretske, 1988; Schleim, 2024). According to the latter, Socrates is in prison because he was sentenced and then faced the punishment. However, the ancient philosopher himself believed in an immortal soul, which may have made his death easier.

The primary science of human behavior has traditionally been psychology. One of its most important schools in the twentieth century, behaviorism, already bore this claim in its name. What was later and still is discussed in philosophy as "eliminative materialism" (e.g. Churchland, 1981) was actually anticipated decades earlier by leading behaviorists such as John B. Watson (1878–1958) and Burrhus F. Skinner (1904–1990). The behaviorists wanted to exclude everything that was not scientific in the sense of objectifiability from psychological science. Particularly problematic from that perspective were assumed internal psychological processes. Introspection, the attempt to explain the psyche from within, was completely rejected by Watson as "mental gymnastics"; consciousness also had no place in science (Watson, 1913/1994).

Skinner considered it particularly problematic to explain behavior causally with mental constructs (Skinner, 1953). For example, sentences such as: "The car driver ignored the stop sign because he *wanted* to be home before the football game started," "The woman stood in the queue with the *intention* of being one of the first to get a concert ticket" or "Kim stayed in the house because of *fear*." The fact that there was no room for free will and other central concepts for what it means to be human in this world of thought was already expressed by Skinner in the title of his collection of essays, which was widely read at the time: *Beyond Freedom and Dignity* (Skinner, 1971).

Instead of internal processes, behaviorists wanted to explain behavior from conditioning through the structures of reward and punishment in the environment. In their opinion, these characteristics could be measured and described objectively, i.e. by observing behavior. This school of thought still exists today, for example in the *Journal of Applied Behavior Analysis* (see also Bördlein, 2022; Vargas, 2020). However, apart from its counter intuitiveness and incompatibility with the prevailing view of human nature, this approach has fundamental theoretical problems:

Already for practical reasons, we do not have sufficient control over the preconditions of people's lives to be able to derive and confirm predictions about individual behaviors from general laws, such as in physics or other natural sciences. And the concept of reward, which is central to behaviorists, can only be defined in a circular way: namely as that which makes a certain behavior more likely, while the explanation of what makes the behavior more likely is in turn its rewarding character (Westmeyer, 1973). Thus, even this psychological school, which so much wanted to be an "objective science," did not meet this standard. We do not need to write a history of psychology here, so for our purposes we can leave it at that. However, we will return to this line of thought at the end of the book.

Cognitive and Neuropsychology

Behaviorism was replaced by the "cognitive revolution"—actually it was a "cognitive resurgence"—in psychology (e.g. Fancher & Rutherford, 2017; Wertheimer & Puente, 2020). Why *different* people in the *same* environment behave differently was ultimately to be explained by internal psychological processes. The boxes connected with arrows in the cognitive models, which stand for these processes, sometimes earned this psychological school the name "boxology." As this in itself did not provide an ontological foundation, i.e. no fundamental answer to the question of *what psychological processes actually are*, many psychologists embraced biology in the twentieth century: This gave rise to biological, evolutionary and neuropsychology.

However, we already saw in Chapter 1 that biology has not provided neuropsychiatry with the desired ontological foundation for over 200 years: To this day, for example, there are no reliable diagnostic biomarkers—not for *a single one* of the hundreds of mental disorders in the DSM diagnostic manual (APA, 2022). It could now be argued that the entities of psychiatry—the classified disorders—are abstract diagnostic constructs composed of simpler mental processes; and that there is an ontological basis for these simpler processes in neurobiology (Schleim, 2022a, 2023a).

But this is also not the case, at least not for the more complex cognitive, emotional and social processes that we identified as central to the legal issues in Chapter 2. The fact that psychological constructs cannot simply be grounded in physiological processes has been challenging neuropsychology for over 100 years (Anderson, 2015; Cacioppo & Tassinary, 1990). For example, experienced experimental psychologists recently concluded: "No one knows what attention is" (Hommel et al., 2019). In comparison, Watson's behaviorist critique of consciousness research at the beginning of the twentieth century is interesting—as it is sobering. Here is an original quote:

> The time seems to have come when psychology must discard all reference to consciousness; when it need no longer delude itself into thinking that it is making mental states the object of observation. We have become so enmeshed in speculative questions concerning the elements of mind, the nature of conscious content […] that I, as an experimental student, feel that something is wrong with our premises and the types of problems which develop from them. There is no longer any guarantee that we all mean the same thing when we use the terms now current in psychology. (Watson, 1913/1994, p. 249)

Exactly 110 years later, a vehement dispute broke out in the now undoubtedly established consciousness research community about whether one of the major accounts of consciousness is "pseudoscience" (Lenharo, 2023a, 2024). This debate goes too far here. But we can note that leading researchers in the field are puzzling over the extent to which consciousness is distinct from attention on the one hand and memory on the other (Koch et al., 2016). Other neuroscientists compared the current theories of consciousness and came to the conclusion that they may not even deserve the status of a "theory" (Signorelli et al., 2021). In addition, it is not possible to judge which of these approaches best explains consciousness because there is still no consensus on what such an explanation should achieve or what "to explain" means here.

Other experienced consciousness researchers distinguished six different *explanatory goals*—such as content of consciousness, emotion, phenomenal properties or self-consciousness—and eight different *functions*—such as attention, metacognition, self or working memory (Northoff & Lamme, 2020). They discussed how eight currently pursued approaches differ from one another in these aspects. And these differences are not just theoretical: Depending on their assumptions, researchers found completely different brain regions to be central to consciousness, namely frontal regions on the one hand versus parietal-temporal networks on the other (Koch et al., 2016; Northoff & Lamme, 2020).

If we take Watson's requirement of 1913 that researchers must agree on the meaning of central concepts as a basic precondition for their work, things look highly problematic for parts of psychology and cognitive neuroscience nowadays. In fact, the lack of conceptual and theoretical integration is partly blamed for the ongoing crisis in psychology (Eronen, 2024; Groeben & Westmeyer, 1981; Hutmacher & Franz, 2024).

The fundamental question of which entities the discipline is dealing with at all does not only apply to psychiatry (Hyman, 2021; Kendler et al., 2011; Schleim, 2022a; Vintiadis, 2024). It also applies to psychology. And for the latter, it should be easier to clarify, since, for example, the central definition of mental disorders from the DSM diagnostic manual refers to disorders of cognition, emotion regulation and behavior that reflect a dysfunction of psychological or biological processes (APA, 2022). That is, psychiatry builds on psychology—and not the other way around. In view of the ongoing crisis in psychology and psychiatry, it is not only the approach of "4E cognition" mentioned at the beginning that is currently receiving more attention. In phenomenological psychology, attempts are also being made to establish the discipline from within itself (e.g. Wendt, 2024). For reasons of space, we cannot go into this approach in more detail; but we do not have to from the pragmatic perspective we are aiming for here.

After all, in practical contexts such as law, psychotherapy or psychiatry, unlike perhaps in science and philosophy, we do not have forever to wait for an answer. As quoted at the beginning of the book from the still authoritative US Supreme Court decision on the use of scientific knowledge in court, *Daubert v. Merrell Dow Pharmaceuticals, Inc.* (1993): "Scientific conclusions are subject to perpetual revision. Law, on the other hand, must resolve disputes finally and quickly." Therefore, the availability of a pragmatic approach is important in the meantime. And we will find that law uses pragmatic categories not by pure chance.

Pragmatism in Psychology and Law

I have argued elsewhere that an approach to describing and understanding humans must fulfill three conditions: It must be useful, coherent and meaningful (Schleim, 2024). *Usefulness* means that it proves itself in practice; *coherence* means that its components fit together as good as possible and contradict each other as little as possible; and *meaning* is about the embeddedness in our cultural tradition and language practice.

Against this background, the question immediately arises as to whether radical approaches such as behaviorism, eliminative materialism or even naturalistic reductionism can be consistently defended at all: After all, these positions imply that our language practice is fundamentally wrong. For the behaviorist, all behavior can ultimately be reduced to reward and punishment, whereby these terms in turn can only be defined in a circular way (Westmeyer, 1973); for the last two positions, the level of physics or at least neurobiology is the essential level of reality. In the languages of these sciences, however, there is no definition of content, meaning and truth.

Strictly speaking, but for principle reasons, no representative of these radical views can express true propositions that make sense without contradicting themselves. Interestingly, this fundamental theoretical problem fits in with the fact that the folk psychological views have persisted in law and morality for centuries. The cases analyzed in more detail in this chapter will also show this more clearly.

The free will debate exemplifies this particularly well. After all, representatives from an eliminationist and reductionist perspective have repeatedly declared our common, pragmatic and normative view of humanity to be scientifically refuted. Some have subsequently demanded a revolution in criminal law. Of course, representatives of such views are also free to advocate an alternative model that is useful, coherent and meaningful. However, it is not enough to simply dismiss self-descriptions as "mental gymnastics" (e.g. Watson, 1913/1994), mental processes as "nothing but brain processes" (e.g. Crick, 1994) or conscious decisions as "illusion" (e.g. Wegner, 2002).

Contrary to what is often claimed, "free will," whatever that may be, is neither a cornerstone of our law nor of our morality. Law and morality generally regard us as minimally rational subjects: We are *persons* who, firstly, can usually distinguish right from wrong in terms of the prevailing normative order and, secondly, have the self-control to act on this insight. Our responsibility is derived from this view, unless there is a specific reason for excuse (Bigenwald & Chambon, 2019; Morse, 2007, 2023; Penney, 2012). Moral philosophers similarly speak of people being *responsive to reasons* (e.g. Hirstein et al., 2018). In German criminal law, the relevant categories are called "capacity for understanding and control" (§§ 17, 20 StGB).

Accordingly, anyone who has a severe mental disability, such as experiencing severe psychosis or intense drug intoxication, is sleepwalking or is being forced to do something with sufficient force is not usually held responsible for their acts. But if a person poses a great danger to others, legal systems usually have an option in alternative to punishment. In such cases, people may be placed in psychiatric care until they are no longer dangerous. After all, without responsibility there is no guilt and no punishment—but democratic constitutional states must also take proportionate steps to reduce known dangers to life and limb. Incidentally, not only actions but also omissions can be morally and legally relevant in this sense. This can be the case, for example, if someone does not help another person in need, even though they could and are aware of the emergency.

In this normative system, "free will" is not a necessary building block. Anyone who does not believe this can read the previous two paragraphs again.

I myself prefer to speak of *volitional processes* or *acts of will* (Schleim, 2024). In my opinion, there is no such thing as "will" that is free or not. Even today, philosophers still find it difficult to explain what a will is supposed to be (e.g. Hieronymi, 2022). In other words: I do not want to reify "will." Instead, I am talking about volitional processes that are typically characterized by certain features, such as planning or deliberation in a meaningful context, taking into account one's preferences and desires. We will see in a moment that this view is amenable to empirical research and that the freedom of our decisions is then not a question of "all or nothing," but of "more or less."

Free Will in the Nineteenth Century

As we saw at the beginning of this section, the question of how to explain our behavior is basically as old as philosophy. The conflict between more scientific and more psychological or humanistic approaches became greater in the course of the modern era, the Enlightenment and modernity. After all, more and more precise knowledge was gained about the physiological basis of our perception and movement. Based on his empirical studies, including dissections of animals, René Descartes (1596–1650) developed the idea of living beings as automata; only in humans there would be an immaterial thinking soul that controls the body through interaction with the pineal gland (Descartes, 1649).

In the nineteenth century, the biologist Thomas H. Huxley (1825–1895) fully transferred the idea of automata to humans and thus founded *epiphenomenalism* (Huxley, 1874). According to this view, consciousness would perceive the processes in the world, but not itself intervene in world events. For Huxley, volition (from the Latin *volo*, I will) was more a feeling or to be understood in the sense of freedom of action, i.e. that one can do what one wants to do:

> We are bound by everything we know of the operations of the nervous system to believe that when a certain molecular change is brought about in the central part of the nervous system, that change, in some way utterly unknown to us, causes that state of consciousness that we term a sensation. [...] Other molecular changes give rise to conditions of pleasure and pain, and to the emotion which in ourselves we call volition. I have no doubt that is the relation between the physical processes of the animal and his mental processes. In this case it follows inevitably that these states of consciousness can have no sort of relation of causation to the motions of the muscles of the body. The volitions of animals will be simply states of emotion which precede their actions. (Huxley, 1874, p. 365)

Huxley illustrated this with the example of a frog that you hold in your hand: If you moved the palm of your hand, it would perform automatic movements to keep its balance. However, these movements are not expressions of volitional processes. If the frog were a philosopher, Huxley mused, it would perhaps think that it was consciously causing its own movements. But that would be a mistake. This discussion about the causal role of consciousness and possible illusions of will would continue into our time, as we will see in more detail in a moment (e.g. Schleim, 2021; Wegner, 2002).

In German-speaking countries, too, there were doubts about free will. A particularly striking example of this is the then much-read physiologist Carl Vogt (1817–1895), who wrote:

> Free will does not exist and with it no responsibility such as morality and the criminal justice system and God knows who else want to impose on us. At no moment are we masters of ourselves, of our reason, of our intellectual powers, any more than we are masters of whether our kidneys should secrete or not secrete. The organism cannot control itself; it is controlled by the law of its material composition. What we think in a moment is the result of the current mood, the current composition of our brain [...]. (Vogt, 1852, pp. 445–446)

Strikingly, Vogt not only considered self-control and free will to be impossible, but also drew far-reaching conclusions for law and morality. This in turn prompted other physiologists and eventually also legal scholars to vehemently contradict the statements of Vogt and other materialists.

Two examples of this are the influential anatomist and rector of the University of Vienna Josef Hyrtl (1810–1894) and the renowned physiologist Emil du Bois-Reymond (1818–1896). The former held that there had been no new arguments for materialism since antiquity, but that the media would gladly take up such sensationalist claims (Hyrtl, 1864/1897). The latter formulated his famous "ignorabimus" that it would *never* be possible to fully explain the basis of consciousness and free will in scientific terms. In doing so, he took up older ideas of the philosopher Gottfried Wilhelm Leibniz (1646–1716) and anticipated David Chalmers' "hard problem of consciousness" (Chalmers, 1995; du Bois-Reymond, 1872; Schleim, 2022b).

It is difficult to draw principled conclusions from this type of argument, though. It largely depends on how optimistic or pessimistic we are about the possibilities of future science. It should be remembered, however, that Karl R. Popper (1902–1994) also spoke of the "promissory materialism" around 100 years after du Bois-Reymond (Popper & Eccles, 1977). By this he meant the habit of representatives of materialist positions to repeatedly hold out the prospect of a scientific solution to problems such as consciousness or free will in the near future. Only recently, Christof Koch, one of the leading consciousness researchers of our time, lost a bet against the philosopher David Chalmers to explain consciousness neurobiologically within 25 years (Lenharo, 2023b).

For our purposes here, it is relevant that there were already responses to far-reaching statements such as Carl Vogt's from legal scholars in the nineteenth century, for example from the then influential criminal law professor Franz von Liszt (1851–1919). He wrote in his authoritative textbook, which ultimately appeared in 25 editions:

> Criminal law therefore does not require the assumption of a causeless self-determination, a freedom of will removed from the causal law, in order to lay its foundations. Rather, it is sufficient to assume, which is not seriously disputed by any side, that all human action is psychically (not mechanical) caused, i.e. determined by ideas, motivated. (von Liszt, 1900, p. 69)

Von Liszt thus established that criminal law neither presupposes an uncaused will or soul, nor does it contradict the idea of causality. Rather, it is important to regard actions as being caused by *psychological* processes. Put differently: The law does not distinguish caused from uncaused actions, but considers the causes and context of a particular action and whether that justifies an excuse or not. Von Liszt's distinction of "mechanical causes" deserves a more in-depth discussion on another occasion. Note, however, that the concept of mechanism is being developed further philosophically and that some approaches include psychological and social causation (e.g. Craver, 2007; Kendler et al., 2011).

As we have seen, it was pointed out repeatedly since at least the nineteenth century that existing criminal law is compatible with psychological explanations of our behavior—and to date, the normative revolution called for by Vogt and others has not occurred. In Sect. 4.2, we will discuss the actual implementation of a "criminal neurolaw" in the Netherlands. But first we will conclude the discussion of the free will debate.

Free Will in the Twentieth and Twenty-First Centuries

So there were already far-reaching statements about free will in the nineteenth century similar to those we know from our own time. We should also not forget Sigmund Freud (1856–1939): He pointed out in the early twentieth century that we have less conscious control over our mental processes than is often assumed (Freud, 1917/1947; Schleim, 2012). He was thinking of sexual impulses on the one hand and the limited control of the unconscious on the other.

One might think that such a discussion would eventually come to a conclusion—at least if it is not only conducted by philosophers, but also involves empirical researchers. As we saw in the previous section, Carl

Vogt declared free will to be an impossibility over 170 years ago and called for far-reaching moral and legal upheavals. Of course, much more knowledge about the body and mental processes is available today than in Vogt's or Huxley's time. But what is striking about the dynamics of today's debate is that it did not begin with the publication of the much-cited Libet experiments in the early 1980s. Rather, the sharp rise coincided with the "Decade of the Brain" and the emergence of neuroethics and neurolaw (Fig. 3.1).

The neuroscientist Benjamin Libet (1916–2007) was above all a pioneer in the field of consciousness research. He dared to take a look inside the "black box," our brain, when behaviorists like Skinner were still in charge. He would later describe how he was unable to find funding for his research and was even advised against it altogether for strategic career reasons (Libet, 2004). We remember that consciousness was not considered a serious scientific topic at the time. Nevertheless, Libet carried out his pioneering experiments with electroencephalography. And even if his test subjects only had to make a small hand movement, this would make him world-famous. However, this happened in a different way than he himself had imagined.

As the title of his book *Mind Time: The Temporal Factor in Consciousness* illustrates (ibid.), he himself was originally concerned neither with free will nor with refuting it, but with investigating the temporal dynamics of conscious processes using neuroscientific methods. The interpretation of the alleged refutation of free will was imposed on his experiments by other researchers. Instead of presenting the decision to make a movement as being determined by unconscious brain processes, Libet and his colleagues drew other conclusions:

> However, accepting our conclusion that spontaneous voluntary acts can be initiated unconsciously, there would remain at least two types of conditions in which conscious control could be operative. (1) There could be a conscious 'veto' that aborts the performance even of the type of 'spontaneous' self-initiated act under study here. This remains possible because reportable conscious intention, even though it appeared distinctly later than onset of [the readiness potential], did appear a substantial time (about 150 to 200 ms) before the beginning of the movement as signaled by the [electromyogram]. Even in our present experiments, subjects have reported that some recallable conscious urges to act were 'aborted' or inhibited before any actual movement occurred [...]. (2) In those voluntary actions that are not 'spontaneous' and quickly performed, that is, in those in which

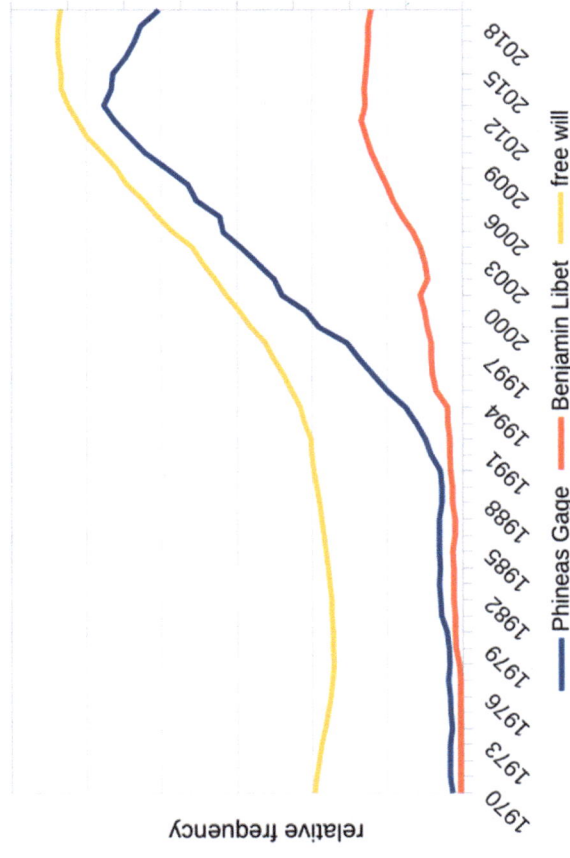

Fig. 3.1 We see the relative frequency of the terms "free will" (yellow), "Benjamin Libet" (red) and "Phineas Gage" (blue) in English-language books over time. There has been a significant increase here since the 1990s. We will come back to Phineas Gage in the next section. As before, the lines are shown on different scales, so they should not be compared directly with each other (yellow: 10^{-6} percent; red and blue: 10^{-8} percent) (*Source* Google Ngram)

conscious deliberation (of whether to act or of what alternative choice of action to take) precedes the act, the possibilities for conscious initiation and control would not be excluded by the present evidence. (Libet et al., 1983, p. 641)

The researchers had thus formulated at least two open possibilities for the conscious control of our decisions: Firstly, there had already been experimental findings at the time that the (supposedly) unconscious readiness potential in the brain also occurred when the movement was not performed at all; this was also confirmed by later repetitions of this and similar experiments (Schultze-Kraft et al., 2016; Trevena & Miller, 2010). Therefore, the readiness potential could obviously not be the (sufficient) cause of the movement. What Libet and colleagues repeatedly described as a conscious "veto" was simply ignored by other researchers who denied free will (Haynes & Eckoldt, 2021; Wegner, 2002). Libet himself criticized this as an error of omission (Libet, 2004).

Secondly, the experimental setup, particularly the spontaneous nature of the movements, did not investigate proper volitional processes as described above at all. In other words: What was researched in the experiment could not be transferred to what is commonly meant by free will, certainly not in a deeper philosophical sense. Using these results to refute free will is thus actually a categorical mistake (i.e. confusing spontaneous movements and willful acts).

The renowned consciousness researcher Anil K. Seth from the University of Sussex later admitted the existence of the veto, but only to immediately relativize it: "Any conscious 'veto', however, is also likely to have identifiable neural precursors [...]" (Seth, 2018, pp. 2–3). In his book on consciousness, published somewhat later, he wrote:

A common interpretation of Libet's experiment is that it 'disproves free will'. Indeed, it is clearly bad news for spooky free will (not that more bad news is needed) because it seems to exclude the possibility that the experience of volition caused the voluntary action. Libet himself was sufficiently worried by this implication, that in what now seems like a desperate rescue attempt, he floated the idea that enough time remained between the moment of the urge and the resulting action for spooky free will to intervene and prevent the action from happening. If there isn't any genuine (i.e. spooky) free will, Libet thought, maybe there's still 'free won't'. This is a cute trick, but of course it doesn't work. Conscious inhibition is no more a little miracle than the original conscious intention. (Seth, 2021)

As I understand it, however, this is a simultaneous change of subject and a straw man argument: The fact that in the 150–200 milliseconds between the conscious decision and the behavior, another cognitive process can intervene and interrupt the execution of the movement is not only an empirical hypothesis made by Libet and colleagues (Libet, 2004; Libet et al., 1983), but has actually been confirmed experimentally (Schultze-Kraft et al., 2016). This has nothing to do with "spooky" free will. Moreover, Seth's account shifts the question of whether a movement is controlled by conscious or unconscious processes and what their temporal sequence is, to speculation about whether movements are caused by brain processes *at all*.

Benjamin Libet and his colleagues simply noted that there is not yet a complete causal explanation for the movement—and the missing variables in the equation can be conscious *or* unconscious processes (see also Dominik et al., 2023; Nestor, 2019). We have already seen in this chapter why the unconscious/conscious distinction is so important for law. It will become clearer in a moment why I am devoting so much attention to this topic. After all, apart from the unfortunately often inadequate presentation of the original data in such discussions, it goes to the heart of the connection between brain research, psychology and our normative practices.

In a variant of the Libet experiment in the brain scanner, the researchers even claimed to be able to predict whether test subjects would press a button on the left or right up to ten seconds before their conscious decision (Soon et al., 2008). The much-cited study—it currently has over 2400 citations on Google Scholar—was immediately presented as a challenge for the legal system both by the researchers and in the media (e.g. Welberg, 2008). It was also claimed that the metaphysical problem of free will is now solved, as we are in reality only following an impersonal causal law that unconsciously determines our decisions (Smith, 2011). We still remember Carl Vogt's similar assertion from the nineteenth century. The results were also specifically referred to the topic of our book:

> Although it is hard to imagine that our decisions might be made subconsciously, these findings have important implications. Can people be held accountable for their actions if they do not become aware of their decisions until after they are made? You decide. (Welberg, 2008, p. 411)

Apart from the fact that this study again examined spontaneous movements and not volitional processes as described above, that two-thirds(!) of the subjects had to be excluded from the experiment because they did not meet the researchers' requirements, in some cases even after data collection, and that the prediction was only slightly above chance level, my central point is that all these experiments, like those of Libet and colleagues and what followed, were about *conscious control processes*. This is even indicated by the brain regions that were associated with the decisions in the latter study, namely parts of the frontal brain and the parietal lobe (Soon et al., 2008). To a certain extent, these are regions *par excellence* when it comes to consciousness. Unconscious influences, on the other hand, are often associated with subcortical structures.

Whichever way you look at it: The human being controlled by the unconscious brain seems to be more a construct of some brain researchers than a result of brain research (Schleim, 2024). As we have seen, law is not based on an abstract notion of free will, but on knowledge of right and wrong in combination with conscious control. In the experiments mentioned here, the test subjects were also forbidden to do everything that constitutes willful acts in the first place: such as planning, deliberation or acting for reasons. If you force people to simulate a kind of random generator, then the result cannot represent more than that.

A fundamental problem also has to do with the aforementioned self-contradiction of materialists and naturalists: How do the researchers in these experiments know which brain processes are supposed to be conscious and which not? As we have seen, we are still far from a complete neuroscientific theory of consciousness. Instead, these researchers expect their test subjects to fix consciousness at a certain point. Everything that took place before then is simply defined as unconscious (Libet et al., 1983; Soon et al., 2008). But this means that in their refutation of free will, the researchers assume on the one hand that people can pinpoint this point in time and on the other hand they subsequently claim that we humans are fundamentally mistaken about the course of our consciousness.

None of this really fits together and certainly does not refute our normative practice. And as it is rightly said: "Extraordinary claims require extraordinary evidence." The general arguments against any possibility of free will are merely *extraordinarily speculative*. But this discussion leads us step by step to the last fundamental point, how the causes of our behavior are related to law and morality and especially the concept of responsibility.

Why Libet's Experiments Don't Refute Free Will

Even forty years later, there's still much ambiguity about how to interpret Libet's experiments. The strongest arguments for why they cannot and do not refute free will are:

- These (and similar) experiments were about spontaneous movements, not willful acts; they particularly lacked planning, deliberation and a meaningful context.
- These (and similar) experiments *required* conscious control instead of refuting its possibility.
- At the reported moment of conscious awareness, the execution of the movement can often still be stopped.
- Because the (allegedly unconscious) readiness potential in the brain also occurs when there is *no* actual movement, it cannot be its (sufficient) cause; this is related to the previous point.
- The original publications by Libet and colleagues don't even mention free will and Libet himself never denied its possibility.
- Neuroscientists recently showed that the measured readiness potential may only be an artifact of the statistical analysis and not even occur in the single trials when subjects make the movement (Schurger et al., 2021; but see Schmidt et al., 2016). This is difficult to decide because of the bad signal-to-noise ratio of EEG.

Determination and Responsibility

We saw above that Carl Vogt in the nineteenth century considered free will to be impossible and criminal law to be fundamentally wrong because all our decisions and actions are determined by natural law. In response to experiments such as Libet's, the argumentation in the twentieth and twenty-first centuries was different: Because our decisions and actions are determined by *unconscious* processes, we have no free will and criminal law is based on a false view of human nature.

Vogt's claims—but also Anil Seth's with his "spooky" free will—were instead directed against a certain philosophical position, namely that of the libertarian (see Roskies, 2006; Schleim, 2024). For the libertarian, our decisions and actions are not completely determined by natural law, but by a speculatively assumed will that transcends the natural order. This point of view can perhaps be taken philosophically, with all the problems that the assumption of such a metaphysical entity "the will" raises (e.g.

Clarke, 2003; Franklin, 2018; Kane, 2009). However, it is not a basic requirement of criminal law.

The argument against the *conscious* control of our behavior is indeed directed against a condition of responsibility in law and morality. However, as I have shown, the experiments mentioned were precisely conscious control tasks. The fact that the instructions encouraged the experimental subjects to behave spontaneously and that those participants who did not behave spontaneously enough were actually excluded does not change this: The subjects had consciously consented in advance, maintained conscious control of their behavior—such as hand movements or button presses—during the experiments and could have stopped at any time. If they had been credibly assured that they were giving life-threatening electric shocks to children, for example, by pressing the buttons, they could have been held responsible.

Accordingly, the neurobiological attack on the foundations of criminal law that has been repeatedly attempted since the nineteenth century presupposes either a confusion of a certain philosophical concept of free will with existing normative practice or a questionable interpretation of experimental data. But it is not only the law that uses this understanding of humans as persons who generally act knowingly and under control: We saw in Chapter 2 that psychologists experimentally measure processes such as cognitive control or impulsivity. These are therefore characteristics that we sometimes have more and sometimes less of and that are more pronounced in some of us than in others. Anyone who wants to undermine criminal law in this way would also have to show that psychologists are actually chasing ghosts when taking such measurements.

Legal Examples

In US criminal law, the causal connection between a certain psychological process, *mens rea* (literally: guilty mind), and a criminal act or omission, the *actus reus* (literally: guilty act), is essential. For example, if John ambushes his former girlfriend's new partner out of revenge and shoots him, this is a homicide. If, by chance, Jackie shoots the man in the same place but a moment earlier because she has a psychosis and believes him to be Satan himself, John could—despite having the same consciousness and behavior—at most be charged with *attempted* murder. After all, you can't murder an already dead person. In this case, therefore, the necessary causal link for the murder offense is missing: It was Jackie, not John, who killed the man (see Dressler, 2015).

The last example shows that the existence or nature of an offense depends not only on the characteristics of the person, but also on the world; law and morality therefore do not only take place in our heads or brains. Let's imagine 999 people driving a car in a hurry and exceeding the speed limit on their way home. But for the thousandth person, a child unexpectedly jumps onto the road and is fatally injured by the car. Even if all of the thousand people subjectively do the same thing, only the thousandth person commits a negligent homicide. In philosophy, this problem is also called the problem of moral luck (e.g. Nagel, 1979).

At best, this unlucky driver could avoid responsibility for the accident if, for example, they had lost control of the vehicle due to a serious impairment of consciousness, such as an epileptic seizure. But even then, the question is not whether the act was caused or uncaused, or whether metaphysical free will intervened in the event or not—but rather the *nature* of the cause. Or to quote a classic court decision of the British House of Lords concerning the homicide of a hitchhiker:

> No act is punishable if it is done involuntarily: and an involuntary act in this context – some people nowadays prefer to speak of it as 'automatism' – means an act which is done by the muscles without any control by the mind such as a spasm, a reflex action or a convulsion; or an act done by a person who is not conscious of what he is doing such as an act done whilst suffering from concussion or whilst sleep-walking.[1]

Again, the question is whether people had control over what they were doing. Criminal law professor Joshua Dressler of Ohio State University explained this in his influential textbook:

> Thus, when D's arm strikes V as the result of an epileptic seizure, we sense that D's body, but not D the person, has caused the impact. In the context of the criminal law, the movement of D's arm is conceptually the same as a tree branch bending in the wind and striking V. When D 'wills' her arm to move, however, we feel that D, and not simply her arm, is responsible for V's injury. Her 'acting self' is implicated. *A* personal, human agency is involved in causing the bodily contact. (Dressler, 2015)

---

[1] Bratty v. Attorney-General, 1963, A.C. 386, United Kingdom House of Lords, online at: https://www.bailii.org/uk/cases/UKHL/1961/3.html.

As early as the 1990s, criminal law professor Stephen J. Morse of the University of Pennsylvania identified it as a "psycholegal mistake" to regard causation in itself as a reason for excuse (Morse, 1994). After the later emergence of neurolaw, he explained:

> For purposes of assessing responsibility, it does not matter whether the cause of the behavior in question is biological, psychological, sociological, or some combination of the three. Adducing a genetic or neurophysio-logical cause does no more work than adducing an environmental cause. The question is always whether the legal criterion for non-responsibility in question is met, however that condition may have been caused. (Morse, 2007, p. 217)

The assessment of criminal responsibility is therefore, as we have seen above, about the *psychological* processes involved:

> As a matter of current, positive law, an agent will be prima facie crimi-nally responsible if the agent acts intentionally and with the appropriate mental state, the mens rea, required by the definition of the offense, such as purpose, knowledge, recklessness, or negligence. Criminal law typically defines an act as an intentional bodily movement performed by an agent whose consciousness is reasonably intact. (ibid., p. 210)

We have already described psychological processes as embodied and embedded at the very beginning of this book. Legal scholars and empir-ical scientists alike have repeatedly pointed out that our behavior is (also) biological (e.g. Greenberg & Bailey, 1994; Hyman, 2021; Turkheimer, 1998). Neither our psychology nor the criteria of criminal law contra-dict this. Therefore, uncovering the neuronal basis of our psychological processes *cannot* in principle refute this normative practice.

Rather, it would be a sensation if brain research were to identify an action as *uncaused*. But instead of changing one's materialistic or naturalistic convictions, one would then probably suspect a mistake in the experimental setup or assume that one would have to search even longer. Nor do we seem to be approaching eliminative materialism; on the contrary, brain researchers are uncovering the structures underlying cognitive control, for example in the dorsolateral prefrontal, anterior cingulate and parietal cortex (e.g. Breukelaar et al., 2017). In other words, our psychological knowledge is underpinned by neurobiology, perhaps refined in individual cases, but not fundamentally replaced.

In this respect, the free will debate repeatedly turned out to be a false alarm, both in the nineteenth century and in the twentieth and twenty-first centuries. The fact that Benjamin Libet and his colleagues saw possibilities for conscious control from the very beginning could have been taken seriously as a warning signal (Libet, 2004; Libet et al., 1983). But then, of course, the conclusion would have been much less sensational: "Human decisions are subject to conscious control even in laboratory experiments, study finds."

The fact that media reports gave contradictory accounts of such experiments, yet prioritized the sensational interpretation, was also a warning sign (Racine et al., 2017). We recall that the anatomist Josef Hyrtl was already aware of this media bias in the nineteenth century (Hyrtl, 1864/ 1897). And if we cannot agree on the significance of an experiment for free will in the long term, then it is probably not a meaningful experiment on this topic at all.

For the following section, in which we deal with the normative consequences of brain injuries, this observation is particularly important: Law and morality assume for our practice of responsibility that we can generally distinguish between right and wrong and act according to this insight. We are perhaps not always, but often enough, minimally rational actors. Incidentally, this does not in principle rule out emotional, irrational or unconscious influences on our decisions and behavior, as several examples have shown. This practice is also useful, coherent and meaningful—which is reflected not least in the history and enduring existence of various legal systems around the world, despite the claims of some scientists and philosophers that they are based on error. The fact that there are controversial borderline cases not only occupies the courts, but also legal and ethical research. However, such borderline cases do not undermine the overall meaning and purpose of the criteria used, just as cases of doubt between the categories of "healthy" and "sick" do not invalidate the general meaning of this distinction (Schleim, 2023a).

3.2   Famous (Ir)Responsible Brains

Toward the end of the 1960s, three North American doctors wrote a letter entitled "Role of brain disease in riots and urban violence," which was published in the *Journal of the American Medical Association* (Mark et al., 1967). The last author, Frank R. Ervin (1926–2015), would a few years later co-author the book *Violence and the Brain* and be praised in his obituary as a "world leader in appreciating the importance of brain disease to disordered behavior of all kinds" and an expert on "multiple presidential advisory committees on problems of violence, crime and delinquency."[2]

The thought of the three doctors was that although many people suffer from poverty, unemployment, living in slums and poor education, not all of them become violent. Couldn't this difference be due to individual differences between people, particularly in their brains? They referred to studies by a "Neuro-Research Foundation" at the time, according to which criminals were more likely to have brain abnormalities (Mark et al., 1967; see also Pustilnik, 2009). These studies focused in particular on the amygdalae, part of the limbic system and located inside the temporal lobes (Fig. 3.2). Even today, these small nuclei in the brain still receive a great deal of attention in the context of neuroscience and law (e.g. Tonnaer et al., 2023).

The American doctors' proposal coincided with the civil rights movement, in which black people fought against discrimination. According to Jonathan M. Metzl, psychiatrist and professor of sociology and medicine at Vanderbilt University in Nashville, the diagnosis of schizophrenia also changed to a "black disease" at that time (Metzl, 2009). In other words, the aggression with which oppressed people fought for equal rights was pathologized neurobiologically or psychiatrically from various sides. Under the heading "Assaultive and belligerent?" there were also advertisements for the "appropriate" psychotropic drugs at the time, such as the antipsychotic haloperidol (ibid.).

However, the proposal, with the participation of Frank R. Ervin, to attribute violent protests to brain abnormalities met with criticism within the medical profession. For example, one critical response noted that the majority of people with brain damage do not become violent and that most aggressive people have no recognizable brain abnormality (Pollack, 1967). The proposal was also criticized as racist by black activists (see Rollins, 2021). Such accusations would haunt Ervin for years to come, and while he acknowledged that the deaths in the 1960s protests were primarily due to police officers and National Guardsmen, he still maintained that 25% of his patients with impulse control problems had a brain disorder (Ervin, 1973).

---

[2] https://www.mcgill.ca/psychiatry/channels/news/obituary-dr-frank-R-ervin-1926-2015-249027.

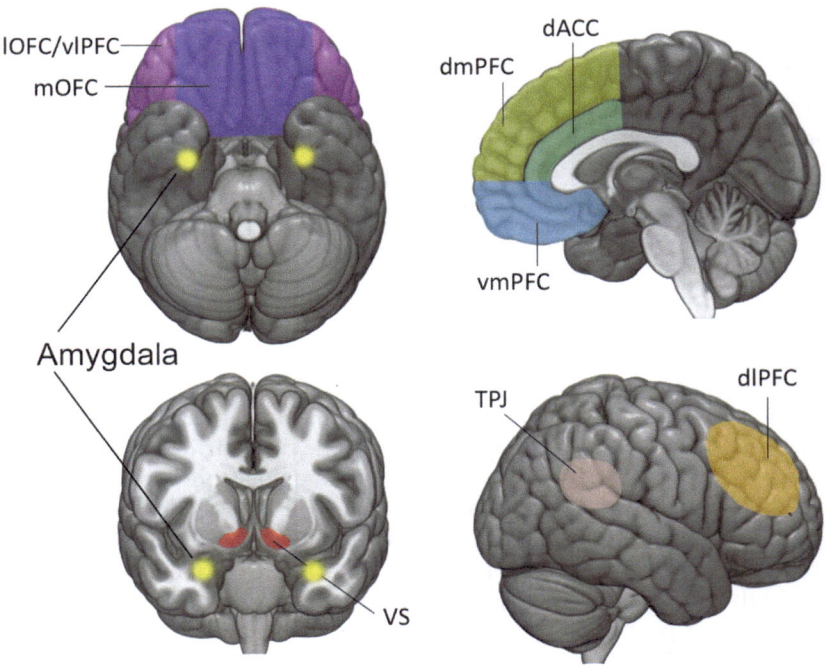

Fig. 3.2 Various brain regions that are of particular importance in this section are marked here. The lower-case letters roughly indicate a direction (d = dorsal; l = lateral; m = medial; v = ventral). The abbreviations in capital letters stand for: ACC = anterior cingulate cortex; PFC = prefrontal cortex; OFC = orbitofrontal cortex; TPJ = temporo-parietal junction; VS = ventral striatum (*Source* Messimeris et al. [2023]. License: CC BY 4.0 [http://creativecommons.org/licenses/by/4.0/])

The discussion at the time, which of course followed on from the biological criminology founded by Cesare Lombroso (1835–1909), is important for us today from the perspective of the theory of science. For in order to be able to make statistically reliable statements about how often violent crimes are associated with brain disorders, we need four sources of information: firstly and secondly, the number of violent criminals with/without such disorders; and thirdly and fourthly, the number of *non*-violent people with/without such disorders. This requires not only a clear understanding of the brain abnormalities being sought; depending on the definition, up to 25% of people could have neuronal norm variants or 3–8% clinically relevant abnormalities (Schleim et al., 2007a). Above all, sufficient representative brain scans of violent criminals and innocent people would have to be available. To date, there has been no breakthrough in this regard. We should keep this limitation in mind when we now look at some specific examples.

The Most Famous Patient

The most famous neurological patient is probably Phineas Gage (1823–1860), the railroad worker who lost part of his frontal brain in an uncontrolled explosion in 1848. As we saw in Fig. 3.1, his case—as well as the topic of free will and the name Benjamin Libet—received much more attention since the "Decade of the Brain." The reconstruction of his brain injury by the neurologist Antonio Damasio and colleagues, who digitized the skull kept in a museum (Damasio et al., 1994), was decisive for this. According to the study, the iron rod with which Gage was trying to secure an explosive charge destroyed his ventromedial prefrontal cortex (vmPFC) on both sides (Fig. 3.2). However, more detailed follow-up studies revealed that probably only the left side of this area was affected (Ratiu et al., 2004; Van Horn et al., 2012). Incidentally, it should be noted that indications such as "ventromedial," the small letters in Fig. 3.2, are only approximate directional indications and do not designate precisely anatomically defined brain structures.

It must have seemed like a miracle to people living in the mid-nineteenth century, and perhaps even to us today, that someone could survive such a serious brain injury. Moreover, Gage not only had to survive the iron bar, but also the poor hygienic conditions of the time, and in the weeks following the accident he actually almost died of an infection (Harlow, 1848). However, the case became of lasting interest due to the (alleged) personality changes attributed to him. In short, Gage was described as having gone from being an exemplary employee to a kind of psychopath: dishonest, abusive, impulsive, immoral and no longer capable of sustained work (e.g. Damasio et al., 1994; Weber et al., 2008). Another doctor, however, with whom Gage spent several weeks, found that "the patient has quite recovered in his faculties of body and mind, with the loss only of the sight of the injured eye" (Bigelow, 1850, p. 14).

Others and I have written repeatedly about the portrayal of Gage's personality and life (e.g. Macmillan & Lena, 2010; Schleim, 2012, 2022c). Already in response to Damasio, a neurosurgeon interested in the history of his discipline drew attention to a particular and possibly very important difference: Harlow, who attested to his patient's strong personality changes, was a convinced phrenologist; Bigelow, on the other hand, who did not notice any peculiarities in Gage's personality, doubted the localizationist models of brain and mind of the time (Barker, 1995). The patient was not subjected to psychological examinations as we know them today. This leaves room for speculation. What can be said with certainty, however, is that a lot of myth-making took place over time and even scientific textbooks and articles often do not accurately reflect the case (Schleim, 2022c). In addition to the issue of personality changes, Gage's later work as a coachman is of particular concern. For this, he not only had to be physically fit, but also be able to deal with horses and passengers—and stick to the timetable.

This is particularly important for the question of rehabilitation. According to many accounts, his (alleged) psychological damage was irreversible. The historical evidence that he was later able to return to permanent work clearly contradicts this. In one publication, I discussed this question under the term "neurodeterminism," even if "neurofatalism" might have been more appropriate (ibid.). Whether someone's personality is permanently changed—and in this sense determined—by a brain injury or whether lost functions can be taken over by other structures is important in many ways: for the person themselves and their therapy, for relatives and for social and legal issues, not least in forensic reports on the dangerousness of an offender. We also keep in mind for the following examples that much depends on the accuracy of the presentation of a case study.

Gage's story is, as I once put it in a popular science article, perhaps "too good to be *not* true" (Schleim, 2023b). The most common version is particularly important for those who take a modular view of the brain. Damasio and some others described the ventromedial prefrontal cortex as the seat of morality, so to speak—or at least of functions that are essential for social and moral behavior (Damasio, 1994; Damasio et al., 1994). This modular view is, of course, more elaborate than what phrenologists assumed in the nineteenth century, but has nevertheless been criticized by some cognitive scientists as "new phrenology" (e.g. Dobbs, 2005; Uttal, 2001). Indeed, the tenability of localizationist thinking is still debatable today (e.g. Cacioppo & Tassinary, 1990; Noble et al., 2024; Schleim & Roiser, 2009). We will return to this question in Chapter 5.

For this section, the important preliminary conclusion is that the lack of a general and robust one-to-one correspondence between brain regions and a particular behavior also limits the forensic validity after brain injury: So you can't just tell from the brain whether someone is going to commit crimes or not, as was the dream of the phrenologists and Lombroso. This is quite analogous to what we noted in Chapter 1 about the lack of biomarkers for diagnosing mental disorders. And for the question of responsibility, as we will see in more detail in the following examples, contextual knowledge about the life and behavior of an affected person is also important.

Phrenologists in the Courtroom

Before we look at more recent cases, however, a brief reference to another example from the nineteenth century is in order. In contrast to Phineas Gage, who to our knowledge never came into conflict with the law, the events of November 1834 in Durham near the city of Portland in the US state of Oregon involved a real violent crime:

Nine-year-old Major Mitchell had lured eight-year-old David Crawford into a wooded area on a day when school was canceled due to the teacher's illness. There he first tried to drown the younger boy in a stream. But when that failed, he tore off his victim's clothes, tied him naked to a tree and abused him for hours, including numerous blows with a stick, which resulted in bloody wounds. Mitchell also mutilated Crawford's genitals with a piece of metal. In the end, the perpetrator made another unsuccessful attempt to drown his now weakened victim, but the water was apparently not to be deep enough (Neal, 1835). Crawford is said to have insulted Mitchell beforehand. This case is interesting for us because the perpetrator was examined from a phrenological point of view at the time. A drawing of his head was also made for this purpose (Fig. 3.3).

Even if the opinions of various phrenologists differed on certain details, they jointly inferred "the character of a cowardly, bloody-minded, able villain" from the shape of the boy's head (Neal, 1835, p. 307). (This could also, mind you, be concluded simply from the well-known facts of the crime.) In particular, Mitchells organ for "Destructiveness" was said to be exceptionally large due to a head injury after a fall in early

Fig. 3.3 According to phrenologists, the drawing of the nine-year-old perpetrator Major Mitchell by the artist E. Seager shows a particularly large organ for "Destructiveness" next to the ear, which is said to have been caused by an earlier head injury (*Source* Neal [1835]. License: public domain)

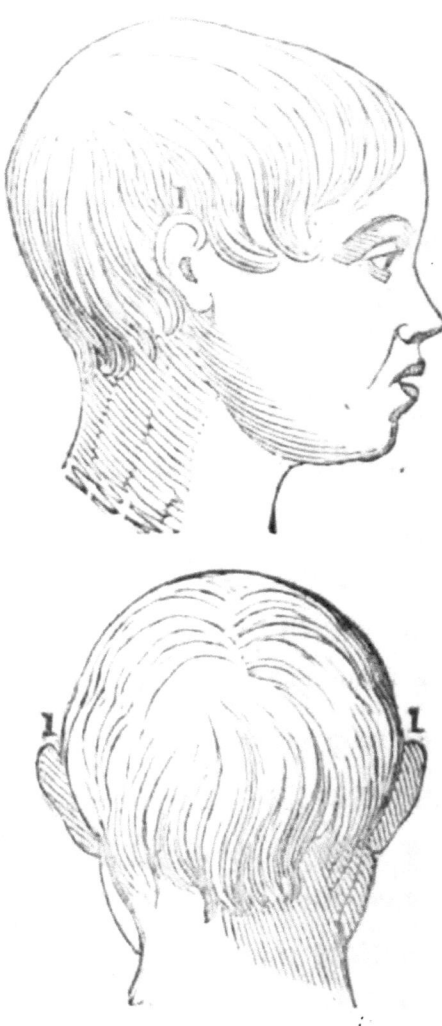

childhood. (I cannot see that on the figure.) However, the court was not convinced by this and the young offender was sentenced to nine years in prison. In particular, it had not been possible to demonstrate that such head injuries regularly lead to such problematic behavior.

We remember what we stated in response to the discussion about the "dangerous brain" in the 1960s: To assess the connection, we would need representative studies of the four groups—violent criminals and innocents each with and without such brain damage—and not just descriptions of individual cases. Nevertheless, the 1834 case is probably the first example of neurolaw in a North American court. Despite the guilty verdict, phrenologists celebrated it as a success:

> We could have wished that the first case for the introduction of phrenology into a court of justice, might be a strong one and prove successful; then would have been afforded an opportunity for a triumphant vindication of its utility, and an augury of its future stupendous influence. [...] Phrenology has been mentioned seriously in a court of justice without provoking laughter. Two most respectable physicians have acknowledged their belief in phrenology, as a science, upon oath; and there were many others here ready, whenever a case might require their help, to submit themselves to further interrogation. (Neal, 1835, p. 309)

We see how the appearance in court here was interpreted as an indication of the public importance and usefulness of the theory. And how eager the proponents of such a brain theory of crime can be to be heard and taken seriously as expert witnesses. Perhaps the latter not only fulfills the desire of some researchers for broad recognition, but also the practical relevance of their research. We recall Nobel laureate Roger Sperry's recommendation in Chapter 1 that neuroscientists should hold out the prospect of solving practical problems in order to present their research as particularly relevant and worthy of support. After these excursions into the nineteenth century, let us turn to more current cases.

Brain Damage and Crime

In addition to Phineas Gage's accident, the case study by the two neurologists Jeffrey M. Burns and Russell H. Swerdlow is still frequently cited in the neurolaw literature (e.g. Swaab & Meynen, 2023). The neurologists from the University of Virginia Hospital reported a case of "acquired pedophilia" due to the brain tumor of a 40-year-old man (Burns & Swerdlow, 2003).

According to the report, the patient began collecting pornographic material in 2000, frequently visiting pornographic sites on the Internet and using the services of prostitutes, which he reportedly had not done before. Much of the pornography is said to have been about children and adolescents. Finally, he approached his pre-pubescent stepdaughter in a sexual manner. The girl turned to her mother after a few weeks, which is how the man's illegal behavior became known. According to the report, a court convicted him of sexual molestation of a child and gave him the choice of either undergoing therapy for "sex addiction" or serving a prison sentence (ibid.).

However, the 40-year-old also attracted negative attention during the treatment: For example, he allegedly asked staff and other participants for sexual favors. Because of this behavior, he was excluded from the therapy and had to start his prison sentence after all. However, on the evening before going to prison, he went to a hospital and complained not only of headaches, but also suicidal thoughts and the fear that he might rape his landlady. At first he received psychiatric care, but due to balance problems he was finally given a neurological examination. He again harassed the clinic staff and urinated on himself, but this did not bother him. An MRI scan finally revealed a large tumor and a cyst in the orbitofrontal and dorsolateral prefrontal cortex (see also Fig. 3.2).

This case study is particularly interesting because the man's problematic behavior disappeared after the tumor was removed. He was finally able to successfully complete the court-ordered therapy and move back into his wife's apartment with his stepdaughter. A little later, however, he began to have headaches and secretly collect pornographic material again. A new examination revealed a return of the tumor, whereupon the patient underwent a second operation. No more problematic behavior is reported after this (Burns & Swerdlow, 2003). We can now apply what we have learned in the long section on causation and responsibility:

First of all, the diagnosis of *acquired* pedophilia—as opposed to one that has always been present—is based on the testimony of the man himself. Since sexual attraction toward children is extremely stigmatized socially, this claim should not be taken as the pure truth. We do not know whether he may have had such inclinations before, but was able to control them better. The keyword "control" leads us to the criteria of minimal rationality identified above, namely knowledge and conscious control. In my opinion, the fact that the man was still able to distinguish right from wrong is shown by his efforts to conceal his problematic behavior over a longer period of time. The fact that he finally sought help at the clinic because he was afraid of committing suicide or further sex crimes also shows a certain degree of both knowledge and control. In contrast, the harassment during his addiction therapy, but also in the neurologic clinic, indicates an increasing loss of control, eventually even over his bladder.

Unfortunately, the neurologists do not report whether the discovery of the tumor—which occurred after the court's sentence—was dealt with again in court and influenced the guilty verdict. In my opinion, this case study certainly shows an important role for neurolaw. However, I think it is an exaggeration to claim that the pedophilia can only be explained by the tumor or that his problematic behavior can only be attributed to it. We can see how contextual knowledge remains important even with such clear findings in the brain. For a neurological interpretation, however, it helps that the orbital/ventromedial regions in the frontal brain are often associated with social behavior and the dorsolateral prefrontal cortex with impulse control (Breukelaar et al., 2017; Messimeris et al., 2023). This case illustrates how the neural findings can support an explanation of the behavior and may well be relevant to assessing the person's responsibility. However, to think that the tumor has turned him into a mindless robot seems implausible to me, given the information available.

No One-to-One Determination

However, the interpretation of such cases is complicated by the fact that damage to a brain region is not *always* associated with criminal behavior. There are also reports that someone with an injury similar to that of Phineas Gage stood out primarily by repeating the same jokes over and over again—or that a man *stopped* his criminal behavior after an unsuccessful suicide attempt with a crossbow and a corresponding injury in the frontal brain (e.g. Mobbs et al., 2007; Schleim, 2011, 2012). The aforementioned thesis of neurodeterminism or neurofatalism has thus been refuted. Family and community support play an important role in healing and reintegration. Gage also wanted to travel to his family as quickly as possible, which, however, the attending physician interpreted as impulsive and childish behavior and was later seen as a sign of his alleged uncontrollability (Harlow, 1848).

The conclusion against neurodeterminism is not "just" theory: Back in the early 1970s, the neurologist and bestselling author Oliver Sacks (1933–2015) reported interesting differences in patients with the same brain disease. When comparing different clinics, he noticed that patients with severe encephalitis lethargica were more active and independent when they were given more freedom and social contact; in a clinic that was more organized like a prison, on the other hand, the patients were more lethargic and also showed worse, Parkinson's-like neurological symptoms (Sacks, 1973/1999).

This illustrates an important explanatory factor: The three doctors who wanted to explain the violent protests of some black people in the 1960s in neurobiological terms held the environment constant (e.g. poverty, slums, poor education, etc.); Sacks, on the other hand, held the neurobiological disease constant (the severe brain inflammation) and compared the effects of different environments (more or less free clinics) on behavior. As with the interpretation of the (alleged) free will experiments, we see again that a reductionist conclusion presupposes a certain perspective-taking by the doctor or researcher. For example, the doctors could have imagined that an aggressive man would have remained peaceful in a different environment. Most likely, even those protesters who became violent under the particular historical conditions were not aggressive for most of their lives. It therefore makes no sense to try to explain brain function or behavior without a specific environment, as I already noted in the foreword.

Once again we see that brain problems must be seen in their psychosocial context. However, the example of sexual harassment in the context of the tumor also shows the fundamental forensic relevance of such findings—and confirms rather than refutes the legal criteria for responsibility. Attempts to explain and possibly even excuse criminal behavior through brain changes date back at least to the early nineteenth century. And they are not always successful.

Normal Evidence in Court

Even after John W. Hinckley Jr. carried out the assassination attempt on then US President Ronald W. Reagan (1911–2004) in March 1981, brain scans were introduced in the court proceedings: Computer tomography was used to show that the would-be assassin had widened sulci in his brain. According to the defense, this feature is more common in people diagnosed with schizophrenia. The judge initially prohibited the showing of the brain images, but then allowed them (Sallet, 1985). Even though an expert witness for the prosecution doubted the validity of these images, his defense's strategy ultimately worked and Hinckley was given mandatory psychiatric confinement instead of being sent to prison. However, the extent to which the psychiatric diagnosis or even the brain scans were decisive for this is still disputed today (e.g. Aggarwal & Jain, 2024; Morse, 2023).

Nevertheless, it is now clear that neuroscientific procedures have found their way into North American court proceedings and will continue to do so. This applies in particular to murder trials and other capital crimes. But this often involves the sentencing phase, after the trial has already resulted in a "guilty" verdict. In this situation, the perpetrator is given extensive freedom to provide exculpatory information. This is why the criteria for scientific evidence are then set lower.

According to empirical studies of hundreds to thousands of such sentences, only a minority of attempts to achieve a reduced sentence using neurobiological methods is successful in practice (Denno, 2015; Farahany, 2015; Greely & Farahany, 2019; Khalid et al., 2024). And even if the success rate then is higher than in trials without neuroscientific material, this difference cannot simply be explained by this; after all, the cases could also have differed in other characteristics, such as generally putting more effort into the defense. Of legal interest, however, is the expectation of some courts nowadays that an appropriate defense also includes a brain examination of the client.

Almost 200 years after the first attempts by phrenologists, we can thus state that neuroscientific procedures have become "normal" in a certain sense, at least in criminal proceedings. But also today's data show an ambiguous association between the presentation of brain abnormalities and, for example, a more lenient sentence. In rare cases, such abnormalities are even used to argue that an offender is particularly dangerous (Denno, 2015). However, in line with our critical discussion, the relationship between a person's brain and behavior in the past, present and future remains complex. Neuroscientific procedures are therefore one piece of evidence among many others and probably not leading to a revolution in the courtroom. From the literature and my own experience, for example as a lecturer at the German Judges' Academy in Trier, I can conclude that judges are highly trained professionals who are generally able to handle new kinds of scientific evidence within the framework of established procedures. I will discuss one particular improvement related to the use of brain scans in Chapter 5.

3.3   SUMMARY

While Chapter 2 focused on general principles of psychological and neural development, we looked more specifically at neurolaw in this chapter. After a brief overview of the topics in this field of research, we discussed some psychological principles and the connection between norms and science. Using the historical and current debate on free will, we examined how the established normative criteria for responsibility relate to the results of research. We saw that the view of persons as minimally rational actors is useful, coherent and meaningful; in particular, the criteria of knowledge of right and wrong and of conscious control are generally supported and not refuted by the scientific findings. Contrary to what is sometimes assumed, causation is therefore not an exclusion criterion for responsibility. Instead, it depends on the specific *type of cause* whether a legally exculpatory condition is present.

On this basis, we were then able to analyze the frequently cited case studies on the connection between brain injuries and criminal behavior in more detail. The importance of examples such as that of Phineas Gage should not be underestimated: Not only did we see them cited with increasing frequency along with the rise of neuroethics and neurolaw (Fig. 3.1). In a recent study of the importance of neuroscientific expertise in the courtroom, Deborah W. Denno, Professor of Law and Director of the Center for Neuroscience and the Law at Fordham School of Law in New York, recently devoted a full twelve pages to the old story again (Denno, 2022). In this chapter, it was shown that statements about Gage's alleged personality changes are questionable and that he was able to lead an independent life despite the massive brain damage; in particular, there is a lack of evidence of any criminal behavior. In this context, it is still often overlooked that people can also have psychological problems as a result of severe traumatic experiences, meaning that possible behavioral abnormalities are not necessarily due to the brain injury (Schleim, 2022c).

The case of sexually assaultive behavior in connection with a large brain tumor, which is also frequently cited, *is* legally relevant. However, the further details show that the perpetrator still retained a certain degree of knowledge and control. So even such extensive brain damage does not necessarily turn a person into a mindless robot. Moreover, there are also examples of people with similar brain damage not exhibiting criminal behavior or even ceasing to do so (e.g. Mobbs et al., 2007; Schleim, 2012).

The fact that the same old cases are cited again and again in the neurolaw literature and that these lead to a more differentiated view on closer examination underpins the usefulness, coherence and meaningfulness of the existing legal criteria; and these are primarily psycho-behavioral (Morse, 2023). A more recent comparison of 21 violent and 20 non-violent criminals in Spain found a large cyst in the brain of *one* participant in the former group, similar to the case study of the American neurologists (Bueso-Izquierdo et al., 2019). However, most of the abnormalities detected were classified as clinically irrelevant. We shall return to this in the concluding chapter. This result also supports the conclusion that criminal behavior cannot simply be explained by brain structures. Accordingly, neuroscientific findings can ideally supplement existing explanations, but not replace them. Whether this also applies to examples that specifically target the brain development of those affected will be analyzed in the next chapter.

REFERENCES

Aggarwal, N. K., & Jain, A. (2024). Neuroethics and neurolaw in forensic neuropsychiatry: A guide for clinicians. *Behavioral Sciences & the Law, 42*(1), 11–19.

Anderson, M. L. (2015). Mining the brain for a new taxonomy of the mind. *Philosophy Compass, 10*(1), 68–77.

APA [American Psychiatric Association]. (2022). *Diagnostic and Statistical Manual of Mental Disorders, fifth edition, Text Revision (DSM-5-TR)*. American Psychiatric Association Publishing.

Barker, F. G. (1995). Phineas among the phrenologists: The American crowbar case and nineteenth-century theories of cerebral localization. *Journal of Neurosurgery, 82*, 672–682.

Batts, S. (2009). Brain lesions and their implications in criminal responsibility. *Behavioral Sciences & the Law, 27*(2), 261–272.

Bigelow, H. J. (1850). Dr. Harlow's case of recovery from the passage of an iron bar through the head. *American Journal of the Medical Sciences, 20*, 13–22.

Bigenwald, A., & Chambon, V. (2019). Criminal responsibility and neuroscience: No revolution yet. *Frontiers in Psychology, 10*, 454562.

Blitz, M. J. (2017). *Searching minds by scanning brains: Neuroscience technology and constitutional privacy protection*. Palgrave Macmillan.

Blitz, M. J., & Bublitz, J. C. (Eds.). (2021). *The law and ethics of freedom of thought, Volume 1: Neuroscience, autonomy, and individual rights*. Palgrave Macmillan.

Bördlein, C. (2022). *Methoden der angewandten Verhaltensanalyse: Eine Einführung*. Kohlhammer.

Breukelaar, I. A., Antees, C., Grieve, S. M., Foster, S. L., Gomes, L., Williams, L. M., & Korgaonkar, M. S. (2017). Cognitive control network anatomy correlates with neurocognitive behavior: A longitudinal study. *Human Brain Mapping, 38*(2), 631–643.

Bueso-Izquierdo, N., Verdejo-Román, J., Martínez-Barbero, J. P., Pérez-Rosillo, M. Á., Pérez-García, M., Hidalgo-Ruzzante, N., & Hart, S. D. (2019). Prevalence and nature of structural brain abnormalities in batterers: A magnetic resonance imaging study. *International Journal of Forensic Mental Health, 18*(3), 220–227.

Burns, J. M., & Swerdlow, R. H. (2003). Right orbitofrontal tumor with pedophilia symptom and constructional apraxia sign. *Archives of Neurology, 60*(3), 437–440.

Cacioppo, J. T., & Tassinary, L. G. (1990). Inferring psychological significance from physiological signals. *American Psychologist, 45*(1), 16–28.

Caruso, G. D. (2024). *Neurolaw*. Cambridge University Press.

Chalmers, D. J. (1995). Facing up to the problem of consciousness. *Journal of Consciousness Studies, 2*(3), 200–219.

Chandler, J. A. (2018). Neurolaw and neuroethics. *Cambridge Quarterly of Healthcare Ethics, 27*(4), 590–598.

Chandler, J. A., Harrel, N., & Potkonjak, T. (2019). Neurolaw today—A systematic review of the recent law and neuroscience literature. *International Journal of Law and Psychiatry, 65*, 101341.

Chorvat, T., & McCabe, K. (2004). The brain and the law. *Philosophical Transactions of the Royal Society B: Biological Sciences, 359*(1451), 1727–1737.

Churchland, P. M. (1981). Eliminative materialism and the propositional attitudes. *The Journal of Philosophy, 78*(2), 67–90.

Clarke, R. (2003). *Libertarian accounts of free will*. Oxford University Press.

Cornet, L. J. M., Bootsman, F., & de Kogel, C. H. (2019). Practical implications of neuroscience in the field of criminal justice: Introduction to the special issue. *Journal of Criminal Justice, 65*, 101570.

Craver, C. F. (2007). *Explaining the brain: Mechanisms and the mosaic unity of neuroscience*. Clarendon Press.

Crick, F. (1994). *The astonishing hypothesis: The scientific search for the soul*. Touchstone Books.

Damasio, A. R. (1994). *Descartes' error: Emotion, reason, and the human brain.* Putnam.

Damasio, H., Grabowski, T., Frank, R., Galaburda, A. M., & Damasio, A. R. (1994). The return of Phineas Gage: Clues about the brain from the skull of a famous patient. *Science, 264*(5162), 1102–1105.

Denno, D. W. (2015). The myth of the double-edged sword: An empirical study of neuroscience evidence in criminal cases. *Boston College Law Review, 56,* 493–551.

Denno, D. W. (2022). How experts have dominated the neuroscience narrative in criminal cases for twelve decades: A warning for the future. *William & Mary Law Review, 63,* 1215–1288.

Descartes, R. (1649). *Les passions de l'âme.* Henry Le Gras.

Dobbs, D. (2005). Fact or phrenology? *Scientific American Mind, 16*(1), 24–31.

Dominik, T., Mele, A., Schurger, A., & Maoz, U. (2023). Libet's legacy: A primer to the neuroscience of volition. *Neuroscience & Biobehavioral Reviews, 157,* 105503.

Dressler, J. (2015). *Understanding criminal law* (7th ed.). Matthew Bender & Company.

Dretske, F. (1988). *Explaining behavior: Reasons in a world of causes.* The MIT Press.

du Bois-Reymond, E. (1872). *Über die Grenzen des Naturerkennens.* von Veit & Co.

Erickson, S. K., & Felthous, A. R. (2009). Introduction to this issue: The neuroscience and psychology of moral decision making and the law. *Behavioral Sciences & the Law, 27,* 119–121.

Eronen, M. I. (2024). Causal complexity and psychological measurement. *Philosophical Psychology,* 1–16.

Ervin, F. R. (1973). Violence and brain disease. *JAMA, 226*(12), 1463–1464.

Fancher, R. E., & Rutherford, A. (2017). *Pioneers of psychology: A history* (5th ed.). W. W. Norton.

Farahany, N. A. (2015). Neuroscience and behavioral genetics in US criminal law: An empirical analysis. *Journal of Law and the Biosciences, 2*(3), 485–509.

Franklin, C. E. (2018). *A minimal libertarianism: Free will and the promise of reduction.* Oxford University Press.

Freud, S. (1917/1947). Eine Schwierigkeit der Psychoanalyse. In S. Freud (Ed.), *Gesammelte Werke* (Vol. XII, pp. 3–12). Imago Publishing.

Goodenough, O. R. (2004). Responsibility and punishment: Whose mind? A response. *Philosophical Transactions of the Royal Society B: Biological Sciences, 359*(1451), 1805–1809.

Greely, H. T., & Farahany, N. A. (2019). Neuroscience and the criminal justice system. *Annual Review of Criminology, 2,* 451–471.

Greenberg, A. S., & Bailey, J. M. (1994). The irrelevance of the medical model of mental illness to law and ethics. *International Journal of Law and Psychiatry, 17*(2), 153–173.

Greene, J., & Cohen, J. (2004). For the law, neuroscience changes nothing and everything. *Philosophical Transactions of the Royal Society B: Biological Sciences, 359*(1451), 1775–1785.

Groeben, N., & Westmeyer, H. (1981). *Kriterien psychologischer Forschung* (zweite Aufl.). Juventa Verlag.

Harlow, J. M. (1848). Passage of an iron rod through the head. *Boston Medical and Surgical Journal, 39,* 389–393.

Haynes, J. D., & Eckoldt, M. (2021). *Fenster ins Gehirn: Wie unsere Gedanken entstehen und wie man sie lesen kann.* Ullstein Buchverlage.

Heinrichs, J. H. (2012). The promises and perils of non-invasive brain stimulation. *International Journal of Law and Psychiatry, 35*(2), 121–129.

Hieronymi, P. (2022). What is a will? In U. Maoz & W. Sinnott-Armstrong (Eds.), *Free will: Philosophers and neuroscientists in conversation* (pp. 13–20). Oxford University Press.

Hirstein, W., Sifferd, K. L., & Fagan, T. K. (2018). *Responsible brains: Neuroscience, law, and human culpability.* MIT Press.

Hommel, B., Chapman, C. S., Cisek, P., Neyedli, H. F., Song, J. H., & Welsh, T. N. (2019). No one knows what attention is. *Attention, Perception, & Psychophysics, 81,* 2288–2303.

Hutmacher, F., & Franz, D. J. (2024). Approaching psychology's current crises by exploring the vagueness of psychological concepts: Recommendations for advancing the discipline. *American Psychologist.*

Huxley, T. H. (1874). On the hypothesis that animals are automata, and its history. *Nature, 253,* 362–366.

Hyman, S. E. (2021). Psychiatric disorders: Grounded in human biology but not natural kinds. *Perspectives in Biology and Medicine, 64*(1), 6–28.

Hyrtl, J. (1864/1897). *Die Materialistische Weltanschauung unserer Zeit.* Braumüller.

Jones, O. D. (2004). Law, evolution and the brain: Applications and open questions. *Philosophical Transactions of the Royal Society B: Biological Sciences, 359*(1451), 1697–1707.

Jones, O. D., Schall, J. D., & Shen, F. X. (Eds.). (2022). *Law and neuroscience* (2nd ed.). Aspen Publishing.

Kane, R. (2009). Libertarianism. *Philosophical Studies, 144,* 35–44.

Kendler, K. S., Zachar, P., & Craver, C. (2011). What kinds of things are psychiatric disorders? *Psychological Medicine, 41*(6), 1143–1150.

Khalid, Z., Lee, R., & Wall, B. W. (2024). The use of neurobiological evidence in sentencing mitigation. *Behavioral Sciences & the Law, 42,* 65–78.

Koch, C., Massimini, M., Boly, M., & Tononi, G. (2016). Neural correlates of consciousness: Progress and problems. *Nature Reviews Neuroscience, 17*(5), 307–321.

Kröber, H. L. (2009). Concepts of intentional control. *Behavioral Sciences & the Law, 27*(2), 209–217.

Lenharo, M. (2023a). Consciousness theory slammed as 'pseudoscience'-sparking uproar. *Nature News.* https://www.nature.com/articles/d41586-023-029 71-1

Lenharo, M. (2023b). Decades-long bet on consciousness ends—And it's philosopher 1, neuroscientist 0. *Nature, 619,* 14–15.

Lenharo, M. (2024). Consciousness: The future of an embattled field. *Nature, 625,* 438–440.

Libet, B. (2004). *Mind time: The temporal factor in consciousness.* Harvard University Press.

Libet, B., Gleason, C. A., Wright, E. W., & Pearl, D. K. (1983). Time of conscious intention to act in relation to onset of cerebral activity (readiness-potential). The unconscious initiation of a freely voluntary act. *Brain: A Journal of Neurology, 106,* 623–642.

Ligthart, S., van Toor, D., Kooijmans, T., Douglas, T., & Meynen, G. (Eds.). (2021). *Neurolaw: Advances in neuroscience, justice & security.* Palgrave Macmillan.

Macmillan, M., & Lena, M. L. (2010). Rehabilitating Phineas Gage. *Neuropsychological Rehabilitation, 20,* 641–658.

Mark, V. H., Sweet, W. H., & Ervin, F. R. (1967). Role of brain disease in riots and urban violence. *JAMA, 201*(11), 895.

Messimeris, D., Levy, R., & Le Bouc, R. (2023). Economic and social values in the brain: Evidence from lesions to the human ventromedial prefrontal cortex. *Frontiers in Neurology, 14,* 1198262.

Metzl, J. M. (2009). *The protest psychosis: How schizophrenia became a black disease.* Beacon Press.

Mobbs, D., Lau, H. C., Jones, O. D., & Frith, C. D. (2007). Law, responsibility, and the brain. *PLoS Biology, 5*(4), 693–700.

Morse, S. J. (1994). Culpability and control. *University of Pennsylvania Law Review, 142,* 1587–1660.

Morse, S. J. (2007). The non-problem of free will in forensic psychiatry and psychology. *Behavioral Sciences & the Law, 25*(2), 203–220.

Morse, S. J. (2023). Neurolaw: Challenges and limits. *Handbook of Clinical Neurology, 197,* 235–250.

Nagel, T. (1979). *Mortal questions.* Cambridge University Press.

Neal, J. (1835). The case of Major Mitchell. *Annals of Phrenology, 2*(3), 303–309.

Nestor, P. G. (2019). In defense of free will: Neuroscience and criminal responsibility. *International Journal of Law and Psychiatry, 65*, 101344.

Noble, S., Curtiss, J., Pessoa, L., & Scheinost, D. (2024). The tip of the iceberg: A call to embrace anti-localizationism in human neuroscience research. *Imaging Neuroscience, 2*, 1–10.

Northoff, G., & Lamme, V. (2020). Neural signs and mechanisms of consciousness: Is there a potential convergence of theories of consciousness in sight? *Neuroscience & Biobehavioral Reviews, 118*, 568–587.

Pardo, M. S., & Patterson, D. (2013). *Minds, brains, and law: The conceptual foundations of law and neuroscience.* Oxford University Press.

Patterson, D. M., & Pardo, M. S. (Eds.). (2016). *Philosophical foundations of law and neuroscience.* Oxford University Press.

Penney, S. (2012). Impulse control and criminal responsibility: Lessons from neuroscience. *International Journal of Law and Psychiatry, 35*(2), 99–103.

Pollack, S. L. (1967). Role of brain disease in riots and urban violence. *JAMA, 202*(7), 663.

Popper, K. R., & Eccles, J. C. (1977). *The self and its brain.* Springer.

Pustilnik, A. C. (2009). Violence on the brain: A critique of neuroscience in criminal law. *Wake Forest Law Review, 44*, 183–237.

Rachul, C., & Zarzeczny, A. (2012). The rise of neuroskepticism. *International Journal of Law and Psychiatry, 35*(2), 77–81.

Racine, E., Nguyen, V., Saigle, V., & Dubljevic, V. (2017). Media portrayal of a landmark neuroscience experiment on free will. *Science and Engineering Ethics, 23*, 989–1007.

Ratiu, P., Talos, I. F., Haker, S., Lieberman, D., & Everett, P. (2004). The tale of Phineas Gage, digitally remastered. *Journal of Neurotrauma, 21*(5), 637–643.

Rollins, O. (2021). *Conviction: The making and unmaking of the violent brain.* Stanford University Press.

Roskies, A. (2006). Neuroscientific challenges to free will and responsibility. *Trends in Cognitive Sciences, 10*(9), 419–423.

Sacks, O. (1973/1999). *Awakenings.* Vintage.

Salerno, J. M., & Bottoms, B. L. (2009). Emotional evidence and jurors' judgments: The promise of neuroscience for informing psychology and law. *Behavioral Sciences & the Law, 27*(2), 273–296.

Sallet, J. B. (1985). After Hinckley: The insanity defense reexamined. *The Yale Law Journal, 94*, 1545–1557.

Schleim, S. (2011). *Die Neurogesellschaft: Wie die Hirnforschung Recht und Moral herausfordert.* Heise.

Schleim, S. (2012). Brains in context in the neurolaw debate: The examples of free will and "dangerous" brains. *International Journal of Law and Psychiatry, 35*(2), 104–111.

Schleim, S. (2021). Neurorights in history: A contemporary review of José MR Delgado's "Physical Control of the Mind" (1969) and Elliot S. Valenstein's "Brain Control" (1973). *Frontiers in Human Neuroscience, 15*, 703308.

Schleim, S. (2022a). Why mental disorders are brain disorders. And why they are not: ADHD and the challenges of heterogeneity and reification. *Frontiers in Psychiatry, 13*, 943049.

Schleim, S. (2022b). Stable consciousness? The "hard problem" historically reconstructed and in perspective of neurophenomenological research on meditation. *Frontiers in Psychology, 13*, 914322.

Schleim, S. (2022c). Neuroscience education begins with good science: Communication about Phineas Gage (1823–1860), one of neurology's most-famous patients, in scientific articles. *Frontiers in Human Neuroscience, 16*, 734174.

Schleim, S. (2023a). *Mental health and enhancement: Substance use and its social implications.* Palgrave Macmillan.

Schleim, S. (2023b). Der Fall Phineas Gage: ein Neuromythos. *Psychologie Heute, 2023*(2), 24–29.

Schleim, S. (2024). *Science and free will: Neurophilosophical controversies and what it means to be human.* Springer.

Schleim, S., & Roiser, J. P. (2009). FMRI in translation: The challenges facing real-world applications. *Frontiers in Human Neuroscience, 3*, 845.

Schleim, S., Spranger, T., Urbach, H., & Walter, H. (2007a). Zufallsfunde in der bildgebenden Hirnforschung. *Nervenheilkunde, 26*(11), 1041–1045.

Schleim, S., Spranger, T. M., & Walter, H. (Eds.). (2007b). *Von der Neuroethik zum Neurorecht?* Vandenhoeck & Ruprecht.

Schmidt, S., Jo, H. G., Wittmann, M., & Hinterberger, T. (2016). 'Catching the waves'—slow cortical potentials as moderator of voluntary action. *Neuroscience & Biobehavioral Reviews, 68*, 639–650.

Schmitz-Luhn, B., Katzenmeier, C., & Woopen, C. (2012). Law and ethics of deep brain stimulation. *International Journal of Law and Psychiatry, 35*(2), 130–136.

Schultze-Kraft, M., Birman, D., Rusconi, M., Allefeld, C., Görgen, K., Dähne, S., et al. (2016). The point of no return in vetoing self-initiated movements. *Proceedings of the National Academy of Sciences, 113*(4), 1080–1085.

Schurger, A., Pak, J., & Roskies, A. L. (2021). What is the readiness potential? *Trends in Cognitive Sciences, 25*(7), 558–570.

Seth, A. K. (2018). Consciousness: The last 50 years (and the next). *Brain and Neuroscience Advances, 2.*https://doi.org/10.1177/2398212818816019

Seth, A. K. (2021). *Being you: A new science of consciousness.* Faber.

Shen, F. X. (2016). The overlooked history of neurolaw. *Fordham Law Review, 85*, 667–695.

Shirtcliff, E. A., Vitacco, M. J., Graf, A. R., Gostisha, A. J., Merz, J. L., & Zahn-Waxler, C. (2009). Neurobiology of empathy and callousness: Implications for the development of antisocial behavior. *Behavioral Sciences & the Law, 27*(2), 137–171.

Signorelli, C. M., Szczotka, J., & Prentner, R. (2021). Explanatory profiles of models of consciousness-towards a systematic classification. *Neuroscience of Consciousness, 2021*(2), niab021.

Skinner, B. F. (1953). *Science and human behavior*. Macmillan.

Skinner, B. F. (1971). *Beyond freedom and dignity*. Knopf.

Smith, K. (2011). Neuroscience vs philosophy: Taking aim at free will. *Nature, 477*, 23–25.

Soon, C. S., Brass, M., Heinze, H. J., & Haynes, J. D. (2008). Unconscious determinants of free decisions in the human brain. *Nature Neuroscience, 11*(5), 543–545.

Spence, S. A., Hunter, M. D., Farrow, T. F., Green, R. D., Leung, D. H., Hughes, C. J., & Ganesan, V. (2004). A cognitive neurobiological account of deception: Evidence from functional neuroimaging. *Philosophical Transactions of the Royal Society B: Biological Sciences, 359*(1451), 1755–1762.

Spranger, T. M. (Ed.). (2012). *International neurolaw: A comparative analysis*. Springer.

Stern, B. H. (2001). Admissibility of neuropsychological testimony after Daubert and Kumho. *NeuroRehabilitation, 16*(2), 93–101.

Swaab, H., & Meynen, G. (2023). Introduction: On brain and crime. *Handbook of Clinical Neurology, 197*, 3–9.

Taylor, J. S. (2001). An overview of neurolaw for the clinician: What every potential witness should know. *NeuroRehabilitation, 16*(2), 69–77.

Taylor, J. S., Harp, J. A., & Elliott, T. (1991). Neuropsychologists and neurolawyers. *Neuropsychology, 5*(4), 293–305.

Tonnaer, F., van Zutphen, L., Raine, A., & Cima, M. (2023). Amygdala connectivity and aggression. *Handbook of Clinical Neurology, 197*, 87–106.

Trevena, J., & Miller, J. (2010). Brain preparation before a voluntary action: Evidence against unconscious movement initiation. *Consciousness and Cognition, 19*(1), 447–456.

Uttal, W. R. (2001). *The new phrenology: The limits of localizing cognitive processes in the brain*. MIT Press.

Turkheimer, E. (1998). Heritability and biological explanation. *Psychological Review, 105*(4), 782–791.

Van Horn, J. D., Irimia, A., Torgerson, C. M., Chambers, M. C., Kikinis, R., & Toga, A. W. (2012). Mapping connectivity damage in the case of Phineas Gage. *PLoS ONE, 7*(5), e37454.

Vargas, J. S. (2020). *Behavior analysis for effective teaching* (3rd ed.). Routledge.

Vincent, N. A. (2015). A compatibilist theory of legal responsibility. *Criminal Law and Philosophy, 9*, 477–498.

Vintiadis, E. (2024). Mental disorders as processes: A more suited metaphysics for psychiatry. *Philosophical Psychology, 37*(2), 487–504.

Vogt, C. (1852). *Bilder aus dem Thierleben.* Literarische Anstalt.

von Liszt, F. (1900). *Lehrbuch des Deutschen Strafrechts* (10th Aufl.). J. Guttentag.

Watson, J. B. (1913/1994). Psychology as the behaviorist views it. *Psychological Review, 101*(2), 248–253.

Weber, S., Habel, U., Amunts, K., & Schneider, F. (2008). Structural brain abnormalities in psychopaths—A review. *Behavioral Sciences & the Law, 26*(1), 7–28.

Wegner, D. M. (2002). *The illusion of conscious will.* MIT Press.

Welberg, L. (2008). Free will? *Nature Reviews Neuroscience, 9*, 410–411.

Wendt, A. N. (2024). *Phenomenological psychology as rigorous science.* Springer.

Wertheimer, M., & Puente, A. E. (2020). *A brief history of psychology* (6th ed.). Routledge.

Westmeyer, H. (1973). *Kritik der psychologischen Unvernunft: Probleme der Psychologie als Wissenschaft.* Kohlhammer.

Zarzeczny, A., & Caulfield, T. (2012). Legal liability and research ethics boards: The case of neuroimaging and incidental findings. *International Journal of Law and Psychiatry, 35*(2), 137–145.

CHAPTER 4

The Developing Brain and the Law

[T]he scientists' authority to enter the policy arena rests largely on the credibility of their research findings. (the developmental psychologists Thomas Grisso and Laurence Steinberg; Grisso & Steinberg, 2005, p. 624)

It is important to note that brain immaturity is not meant to remove responsibility for decision making. It describes a predisposition but does not determine behavior. (the developmental neuroscientists Beatriz Luna and Catherine Wright; Luna & Wright, 2016, p. 109)

We have laid important foundations in the previous chapters: Chapter 1 focused specifically on biological psychiatry and the lack of diagnostic biomarkers for mental disorders. In Chapter 2, we looked at the psychological and brain development of young people and found that a term such as "adolescence" needs to be seen in a historical and cultural context and that professionals differ in how they draw the age boundary. Chapter 3 focused on the practice of moral and legal responsibility; in discussing free will and brain lesions, we saw how scientific findings relate to these normative concepts.

With these basics in mind, we can move straight into the discussion of the so-called adolescent brain and its possible significance for criminal law in this chapter: In chronological order, we should first review the discussion of the *Roper v. Simmons* decision of the US Supreme Court

© The Author(s) 2025

S. Schleim, *Brain Development and the Law*, Palgrave Studies in Law, Neuroscience, and Human Behavior,

https://doi.org/10.1007/978-3-031-72362-9_4

from 2005. To introduce the topic, however, I will start with the Dutch "criminal neurolaw" introduced in 2014, which I have already dealt with before (Schleim, 2019, 2020, 2024a).

This example is particularly and, to my knowledge, still unique in how strongly the legislature justified a change in the law based on causal links between brain development and criminal behavior. We then delve deeper into the discussion of US case law, which continues to this day. These two examples from the Netherlands and North America are primarily concerned with the connection between psychological or brain development and criminal responsibility. It is of particular importance that criminal behavior, for example according to data from the US Federal Bureau of Investigation, often begins between the ages of eight and 14, peaks between the ages of 15 and 19 and then declines again between the ages of 20 and 29 (Casey et al., 2022; Moffitt, 2018; Steinberg, 2013). The Dutch figures show a similar pattern, but here the peak is reached slightly later, at 17–20 years of age (van der Laan et al., 2021a). The final example from Germany instead relates more to drug policy and public health.

4.1   The "Adolescent Brain" in the Netherlands

The Dutch example takes us back to the parliamentary elections in the summer of 2010. At the time, several parties were campaigning for more security. The focus was primarily on the criminality of young people and young adults, for both individual acts and gang crime.[1] After the cabinet was formed, the State Secretary for Security and Justice explained this intention in his letter of June 25, 2011.[2] The reason for a special approach to this age group was that they would make up almost 30% of all suspects. Referring to a then newly published expert's report by the Council for Criminal Justice and Youth Protection (RSJ, 2011), a public advisory body formally independent of the government, the Secretary of State explained that the problematic behavior of this age group was linked to certain psychological factors:

---

[1] The following paragraphs in this section are mostly revised translations from Schleim (2019, 2024a).

[2] Parliamentary paper 28 741 No. 17 of June 25, 2011,

Research shows that many psychological functions that are important for the development of socially desirable behavior are not fully developed until after the age of 20. These include the inhibition of impulses, the monitoring and weighing up of long-term consequences, the regulation of emotions and the development of empathy. As these functions are not yet fully developed in adolescents, transgressive behavior and criminality occur relatively frequently, especially in adolescents.[3]

In the explanatory memorandum of December 13, 2012, the State Secretary elaborated this in more detail.[4] Right at the beginning, this was based in detail on a psychological-neurobiological argument. It is therefore instructive for our analysis to quote this fully:

Juvenile criminals already appear to be strongly represented in the crime figures. This prompts me to consider this group separately in the criminal justice system. Recent scientific findings on the development of important brain functions during adolescence support the intention to take a separate approach to adolescents. These findings stem from developmental psychology and are confirmed by recent neurobiological research. They were briefly and succinctly described by the Council for Criminal Justice and Youth Protection in its most recent report with proposals for the design of juvenile (procedural) law. **The fact that specific risk behavior occurs between the ages of 15 and 23 can be attributed in part to the incomplete development of important brain functions.** The main thing that science teaches on this topic is that the mental development of adolescents does not stop at the age of 18, but that important developments only take place after this age. **Incomplete emotional, social, moral and intellectual development is one of the reasons why a large proportion of (juvenile) delinquency occurs during adolescence, but also before the age of 23.** This includes the ability to inhibit impulses (inhibition) and the ability to suppress disturbing impulses and associations (interference). When making risky decisions, the influence of peers appears to be particularly strong up to the age of 20. After that, young people are more capable of making independent decisions in risky situations due to their greater autonomy. The ability to see the long-term consequences of actions and adapt one's own behavior accordingly also appears to develop only after the age of 20. The same applies to the ability to regulate emotions and the development of empathy in young adults. New research into the

---

[3] Parliamentary paper 28 741 No. 17 of June 25, 2011, p. 2.

[4] Parliamentary paper 33 498 No. 3 of December 13, 2012.

functioning of the brain using imaging techniques reportedly shows **that adolescents are guided more than adults by parts of the brain that respond to immediate rewards.**[5]

This quote contains all references to brains or brain parts in the explanatory memorandum. It is the central point at which neuroscience is addressed. In addition, in the places I have highlighted in bold letters, a causal relationship between psychological or brain development and problem behavior is asserted, switching back and forth between the two areas—mind and brain—as if they were one and the same thing. These claims were based on the RSJ report mentioned above. In its section 2.3, entitled "The biological and psychological development of the young person," the scientific research was summarized as follows:

The physical development of normally gifted adolescents shows different stages of development in different areas, which do not always run parallel to each other. For example, the growth in height of adolescents continues until around the age of 16 to 18. The gray matter of the brain develops earlier: there is a growth spurt in the brain until the age of 14 to 15. It is not until around the age of 25 that functions such as planning and flexibility mature in the prefrontal cortex. Other psychological functions such as the inhibition of impulses ("inhibition") and the suppression of disturbing impulses and associations ("interference") also only begin to develop more strongly from the age of 20. **This means that certain risk behaviors that are common in adolescents between the ages of 15 and 23 are partly due to the incomplete development of certain key brain functions.** The most significant development of these brain functions does not take place until after the age of 20. Up to this age, other people, especially peers, still have the greatest influence on risky decisions. It is only after this age that adolescents are able to make their own decisions in risky situations based on greater autonomy. Although adolescents see warning signals when they encounter danger, they do not yet experience them properly and often let the opportunity to stop pass by. **Research into brain function using imaging techniques shows that adolescents are still mainly controlled by the brain nucleus that responds to immediate rewards, the nucleus accumbens, while the brains of people over 25 show greater activity in the amygdala and prefrontal cortex. As a result, the latter group tends to pay more attention to the long-term consequences of dangerous situations.** In the area of emotion regulation,

[5] Ibid., pp. 12–13; my emphasis.

too, it can be seen that the greatest changes do not occur until adulthood and not during adolescence. **It is only when the prefrontal cortex has matured that young people are better able to regulate their emotions than before this time.** Empathy, which plays an important role in the development of norm-compliant behavior, also only takes on a solid form between the ages of 20 and 25. From this it can be concluded that the development of adolescents is generally incomplete until the age of 23. (RSJ, 2011, p. 18; my emphasis)

Once again, I have highlighted in bold the assertions of a causal relationship between the brain, psyche and behavior. Here, mind and brain development are linked even more clearly than in the quote before. The neuroscientific studies cited by the RSJ to support its claims are N. E. Adleman and colleagues (2002), B. J. Casey and colleagues (2005) and T. Paus and colleagues (2001). The remaining three sources are psychological or criminological in nature. I will discuss the neuroscience publications in more detail below. But first, I would like to draw an important conclusion for our purposes, because the bill entered into force on April 1, 2014. This is currently Article 77c, paragraph 1 of the Dutch Criminal Code:

With respect to the young adult who has attained the age of eighteen but not yet twenty-three years at the time of the commission of the offense, the court may, if it finds grounds for doing so in the personality of the offender or in the circumstances in which the offense was committed, dispense justice in accordance with Articles 77g to 77hh.

The articles referred to herein contain the penalties and measures of juvenile criminal law, such as community service, youth detention or long-term placement in an institution for juveniles. These were usually applied up to the age of 17 and sometimes up to the age of 21. Contrary to what the frequently used name "adolescent criminal law" might suggest, the Netherlands has not created a separate criminal law for this age group. Instead, the aforementioned provisions for minors can now be applied to young adults up to and including the age of 22. This is decided by the court at the request of the public prosecutor's office. The Dutch Ministry of Justice's information brochure on the subject states:

> Recent scientific studies on brain development show that some young adults are only able to understand the consequences of their actions and take sufficient account of the effects on others years after the age of 18.[6]

Here, a psychological-behavioral finding is attributed to the brain. In the rules of juvenile criminal law, the pedagogical goal comes before punishment. This is because it is assumed that people in this age group are still more malleable and can therefore change problem behavior more easily. The courts provided the following information:

> Purpose of adolescent criminal law: The stage of development of a defendant over the age of 17 can be a reason for the application of juvenile criminal law. Juvenile criminal law focuses on the best interests of the young person and emphasizes an educational approach, whereas general criminal law focuses on retribution. **The idea is that a defendant's behavior can be adjusted as much as possible while their brain is still developing.** In this way, the likelihood of reoffending is minimized.[7]

In other words, the judges had also adopted the neuroscientific ideas contained in the explanatory memorandum to the law and the RSJ's expert opinion. Today they say in addition:

> The brain of (young) people is usually not fully developed until the age of around 24. Not everyone develops in the same way and at the same speed. And therefore the same punishment or measure does not suit everyone. A judge takes into account the suspect and their stage of development and imposes a sentence that best suits them.[8]

The Dutch criminal law thus takes a more individual approach up to the age of 22: From the age of 12 to 15, the provisions for juveniles are applied; from the age of 16 and 17, as a rule, the provisions for juveniles are applied, but those for adults can be applied instead; from the age of 18 to 22, as a rule, those for adults are applied, but the provisions for

---

[6] "Adolescent Criminal Law: Aanpak met perspectief" of February 2014, p. 5. Dutch Ministry of Security and Justice, online: https://www.wodc.nl/documenten/brochures/2014/02/27/adolescentenstrafrecht---brochure.

[7] https://www.rechtspraak.nl/Organisatie-en-contact/Rechtsgebieden/Strafrecht/Paginas/Adolescentenstrafrecht.aspx (my emphasis).

[8] Ibid.

juveniles can be applied instead; and from the age of 23, adult criminal law is applied. The practical implementation of these rules will be discussed below.[9]

At this point, I consider two conclusions to be important: Firstly, according to my analysis, the Netherlands has a genuine "criminal neuro-law" since April 1, 2014, in the sense that the legislative initiative and justification are largely based on neuroscientific findings. Secondly, however, we see a certain inconsistency in the age limits here: The RSJ's report—in line with our findings from Chapter 2—already refers to neuronal development processes that are still progressing at the ages of 23, 24 and 25.

The Policy Advice Division of the Dutch Council of State, the highest administrative court, also noted this inconsistency in an expert's report.[10] In the interests of consistency, the upper limit for the application of juvenile criminal law should therefore be raised to 24 years. In addition, precisely because of the ongoing development into the early 20s, juvenile criminal law should be applied in principle and adult criminal law in exceptional cases. In the event of practical problems, this could be seen as a transitional rule on the way to genuine adolescent criminal law, according to the state councilors. However, the legislator stuck with the 22 years and the priority of adult criminal law from the age of 18—while the courts, as we saw in the last quote, currently cite the age of 24 as the approximate end of brain development.

We will keep this inconsistency in mind when we take a closer look at the neuroscientific studies quoted in support of the legislative initiative in the next section. It will also play a role afterward, when we discuss the situation in the USA. The attentive reader may still be wondering about the political sense of these measures. In other words, during the election campaign, citizens were promised more security. What does this have to do with the new rules? It was hoped that a more individualized approach would reduce recidivism rates among young offenders. We'll come back

---

[9] On the internet, but also in some scientific papers, it is incorrectly claimed that the new regulation applies up to the age of 23. These errors arise from the Dutch language usage, which strictly distinguishes a statement such as "18 to 23 years" from "18 to 23 years inclusive". As we have seen, the law refers to attaining the age of 23. That is the day on which you turn 23. So Article 77c applies to the age from 18 to 22 inclusive.

[10] Parliamentary paper 33 498 No. 4 of December 13, 2012.

to this as well. And apart from what we have discussed here, the penalties for some offenses have been increased.

Significance of the Studies

It still remains to be answered to which extent the substantive argumentation of the legislative initiative is scientifically justified, i.e. whether the age limit of 22 years can indeed be derived from the studies cited. This concerns the three sources already mentioned, which we will now analyze in more detail (Adleman et al., 2002; Casey et al., 2005; Paus et al., 2001). But it is noticeable already at first glance that this was a rather superficial and not very up-to-date selection, even for a report from 2011.

Adleman and colleagues used the Stroop effect, which has been studied in psychology for almost 100 years and involves suppressing a premature, impulsive reaction. The test subjects had to name the color of a word shown to them. This is easier in the *congruent* version, for example when the word "blue" is shown in blue instead of, say, yellow letters. When word and color are *incongruent*, subjects generally take longer and make more mistakes. In an fMRI scanner, the researchers examined three different age groups, namely 7–11 years olds ("children," $N = 8$), 12–16 years olds ("adolescents," $N = 11$) and 18–22 years olds ("young adults," $N = 11$). It is striking that although the total group of $N = 30$ is normal for such an fMRI experiment, it is far too small to draw general conclusions. Furthermore, for the purposes of the new provisions for adolescents, the comparison with people *over* the age of 22 years of age would have been interesting. Therefore, this study cannot in principle support the upper limit of the law.

The key findings in relation to the age groups mentioned are that young adults showed greater activation than adolescents in three regions of the frontal lobe, namely the anterior cingulate cortex, the left middle frontal gyrus and the left superior frontal gyrus (Adleman et al., 2002). Comparable differences were also found between young adults and children, but not between adolescents and children. In the latter two groups, there were only significant differences in the parietal lobe. On the one hand, this suggests that functions in the frontal lobe that are considered important for cognitive control gradually increase with age. On the other hand, the *differences* found between adolescents and young adults do *not* support the central idea of the new law. After all, these groups are to be treated *equally* under it.

The paper by Casey and colleagues is a review that summarizes various studies on human brain development. In terms of age limits, however, they were rather vague or stated that the brain volume of a six-year-old already accounts for 90% of the adult brain. Furthermore, the sensorimotor, associative and prefrontal regions of the brain are largely developed between the ages of six and 16, they wrote. According to the researchers, the growth of the synapses and dendrites of the nerve cells in the prefrontal lobe is more or less complete by the age of 16 (Casey et al., 2005). Again, this is no reason to treat adolescents and young adults differently in legal terms.

Incidentally, the research of Betty J. Casey from Yale University already played a role in Chapter 2 and we will take a closer look at it in a moment when discussing the US example; the psychologist and neuroscientist is considered one of the leading experts in the field of the "adolescent brain" (Casey et al., 2008). In the 2005 review, the researchers also drew heavily on an essay that called on both scientists and journalists to provide a more accurate portrayal of human development in the media (Thompson & Nelson, 2001).

Regarding the capabilities of brain imaging research, Casey and colleagues wrote in general: "Current non-invasive neuroimaging methods do not have the resolution to delineate which of these processes underlies observed developmental changes beyond gray and white matter subcomponents" (Casey et al., 2005, p. 105). Thus, these methods are not yet specific enough to determine exactly what is changing in the growing brain. Although this does not relate to the justification for the Dutch law, it is interesting to note that Betty Casey published a paper around ten years later with the provocative title "Rewiring juvenile justice." It dealt with the question of the extent to which neuroscience has influenced the law (Cohen & Casey, 2014). As evidence, a study led by her found differences in the brains of 13- to 17-year-olds and 18- to 27-year-olds in impulsive behavior, specifically in emotion-processing limbic and controlling prefrontal brain regions (Dreyfuss et al., 2014). But such *differences* do not justify *equal* treatment of minors and young adults, just as before.

The third and final study, by Paus and colleagues, is another review that summarized studies on human brain development. These researchers presented several analyses with two main conclusions: Firstly, there appear to be developmental processes that continue until the age of 30—data from older people was not included. The change gets smaller and smaller with increasing age. This means that the differences between five- and ten-year-olds, for example, are greater than those between 25- and 30-year-olds. Secondly, the variability within an age group is so great that some six-year-olds have a larger volume in certain structures than some 16-year-olds. This can be seen, for example, in the corpus callosum, which connects the left and right hemisphere (Paus et al., 2001). A continuous but gradually decreasing rate of brain development does not support a hard normative differentiation between children, adolescents, young and older adults, either. However, the large differences within an age group fit in with the approach of the juvenile justice system to take greater account of the individual and not to treat all offenders the same simply on the basis of their age.

At the end of this section, we must conclude that the handling of the neuroscientific studies in the expert's report of the RSJ (2011) and the derived legislative justification for the Dutch "adolescent criminal law" is questionable. Even though the research findings discussed here are older, they are still in line with our findings in Chapter 2. Nevertheless, the Dutch regulations on dealing with defendants and offenders up to 22 years of age have been in force since April 1, 2014, and remain so. This raises the question of how these rules work in practice. We will now look at this in the next and final section on this topic.

Practical Application and Evaluation

As we have seen, the court decides, usually at the request of the public prosecutor, whether the more lenient rules of juvenile criminal law should be applied to a young adult. The law states that "if [the court] finds grounds for doing so in the personality of the offender or in the circumstances in which the offense was committed" (Article 77c, paragraph 1 of the Dutch Criminal Code). According to an evaluation of the amended law, psychosocial circumstances are decisive here: Does the defendant have a mental disability or mental disorder, can they be influenced by their upbringing, how do they function at school, at home, do they live independently, have a steady job? By contrast, a hardened, closed attitude would have a negative impact (van der Laan et al., 2021b). But prosecutors would tend to be intuitive in their respective applications, which could pose a problem for transparency and equality before the law. Other researchers found that, in practice, the seriousness of the crimes made the decision in favor of juvenile justice less likely (Hopman & de Vocht, 2019). This criterion is also not in the spirit of the law.

To improve the application of these measures, another group of researchers identified four dimensions for assessing individual development based on the research literature and the assessments of 19 experts from academia and practice (Spanjaard et al., 2020). These dimensions are: (1) cognitive and adaptive skills (e.g. academic performance, deliberating the consequences of one's actions), (2) social skills (e.g. verbal expression, friendships, vulnerability to the influence of others), (3) moral development (e.g. empathy, guilt) and (4) self-control (e.g. emotion regulation, risk-taking behavior). It goes without saying that the brain plays an important role in all of these skills. In practice, however, they must be derived from psychological tests and a person's known behavior. Someone who has successfully completed school, maintains many friendships, has a steady job and has carried out a criminal act professionally is then likely to be judged differently from someone who is socially isolated, achieves nothing in their life and commits a crime impulsively.

In the meantime, the changes brought about by the legislative initiative have been evaluated in various ways: The most obvious is the increase in the use of the provisions for juveniles from only 0.6 to around 6% of criminal cases with young adults (van der Laan et al., 2021b). However, this mainly affects the group of 18- to 20-year-olds, who were not targeted by the increase to 22 years and for whom this option had already existed since 1965 (van der Laan et al., 2021a). It seems that the public prosecutors and the criminal courts have been made more aware of this possibility by the increased age limit. For comparison: In Germany, around two-thirds of 18- to 20-year-olds were sentenced under juvenile criminal law and only one-third under adult criminal law (Matthews et al., 2018).

The question of whether the more rehabilitative measures will lead to fewer crimes in the long term is more complex to answer. First of all, the number of criminal proceedings against 18- to 22-year-olds had already been falling since the beginning of 2012, i.e. even before the legislative initiative came into force (van der Laan et al., 2021a). With regard to recurring criminality after a conviction, there was no discernible difference in prison sentences without parole between sentences under juvenile or adult criminal law (Prop et al., 2021). An improvement could only be shown indirectly if the offenders were more likely to retain their home, work and social relationships as a result of a measure under juvenile criminal law: "Entering into social ties that require independence is associated with lower recidivism rates regardless of the sanction system applied" (ibid., p. 7). This tends to confirm the idea formulated in Chapter 2 that the problem behavior of adolescents is at least partly a reaction to how society treats them.

However, according to the logic for the application of *juvenile* criminal law in the Netherlands, those who lead a relatively independent life before a conviction would be more likely to be sentenced as an *adult*. The evaluation is complicated by the fact that, as we have seen in this section, the new rules were initially applied intuitively rather than systematically. Interestingly, recent but still preliminary data showed a tendency toward less criminal recidivism after the application of juvenile criminal law (Prop et al., 2021). A practical problem, however, is the additional effort associated with the required psychosocial assessments, which are not equally available in all jurisdictions (van der Laan et al., 2021a).

For our purposes, two observations are important, the first of a theoretical and the second of a practical nature: Firstly, the question of the added value of neuroscience arises at the *justification level* of the legislative initiative. The fact that adolescents and young adults represent a special group from a criminological point of view was already known from the behavioral data, such as the conspicuously high number of criminal offenses. In any case, the neuroscientific studies cited in the relevant expert's report and then in the explanatory memorandum to the law did not match the age limit of 22 years. We will come back to this in the next section.

Secondly, on a *practical level* arises the question of what the various interest groups in criminal law gain from the brain hypotheses of development. We have already seen that the Dutch Ministry of Justice and the courts also refer to the brain to justify the provisions; the same could be shown for the legal profession and expert witness associations. But what use is this if there is no ready-made brain test for individual development and psychosocial criteria have to be used instead? Perhaps the brain or the "neuro" prefix is being used here as a cipher for special scientific credibility. We will return to these questions in the final chapter. For now, we will discuss the ongoing debate on age limits and criminal liability in the USA.

4.2   THE "RESPONSIBLE BRAIN" IN THE USA

In the brief outline of the history of law and neuroscience in Chapter 3, we saw that the term "neurolaw" was originally coined in connection with the treatment of neurological damage in court. However, according to the previously cited content and quantitative analysis by Chandler and colleagues, the neurolaw literature deals primarily with criminal law issues, namely in 34% of publications (Chandler et al., 2019). And within this topic area, the criminal responsibility of adolescents and young adults plays the largest role. This is due to the decision of the US Supreme Court in the case of *Roper v. Simmons* (2005). This came at the time of the emergence of neurolaw. In this section, we analyze the role of theories on brain development in this process.

The case began with a serious crime in the US state of Missouri in 1993. 17-year-old Christopher Simmons wanted to commit burglary and murder. He involved a 15- and a 16-year-old friend in his plans, but the latter left the gang before any crimes were committed. The two remaining teenagers weighed up various targets and finally broke into the house of a 46-year-old woman at night. They tied up and brutally abducted their victim in a van and threw the woman off a bridge into the Meramec River. Due to the restraints, she had no chance of surviving and drowned fully conscious.

Simmons bragged about his crime and was arrested by the police the very next day. After initially denying everything, he later confessed and re-enacted the course of events. A jury found Simmons guilty of murder and recommended the death penalty, which the court then imposed. Incidentally, the accomplice, who was 15 years old when the crime was committed, received a life sentence without the chance of parole. Simmons' case went through the courts and finally ended up in the Supreme Court. The justices then dealt with the question of whether the decision of *Stanford v. Kentucky* from the late 1980s was still valid: that the death sentence can be imposed from the age of 16.

The legal issue to be resolved by the Court in *Roper* was therefore whether the death penalty for offenders aged 16 or 17 at the time of the crime is in line with the Eighth Amendment to the US Constitution. This prohibits "cruel and unusual" punishments. A great deal has been published on the constitutional aspects in recent decades (e.g. Flanders, 2023; Meltsner, 1973/2011; Steiker & Steiker, 2016). The fact that there

Table 4.1 Constitutional decisions of the US Supreme Court on the most severe sentences in which the offender's psychological development played a role

Court decision	Vote	Conclusion
1988: *Thompson v. Oklahoma*	5:3	Death penalty under 16 years unconstitutional
1989: *Stanford v. Kentucky*	5:4	Death penalty from the age of 16 constitutional
2003: *Atkins v. Virginia*	6:3	Death penalty for people with intellectual disabilities unconstitutional
2005: *Roper v. Simmons*	5:4	Death penalty under 18 years unconstitutional
2010: *Graham v. Florida*	6:3	Life sentence without prospect of parole under 18 years only constitutional in homicide cases
2012: *Miller v. Alabama; Jackson v. Hobbs*	5:4	Mandatory life sentence without prospect of parole under 18 years unconstitutional
2016: *Montgomery v. Louisiana*	6:3	*Miller* also applies retroactively
2021: *Jones v. Mississippi*	6:3	No separate investigation is required to assess the incorrigibility of an offender

are different perspectives on this is shown by the narrow majority of 5:4 votes in the Supreme Court's decision.

In line with the theme of this book, we focus below on the question of what role science, and brain research in particular, played in this process. This analysis is made more interesting by the fact that after *Roper* there have been several subsequent decisions on the constitutionality of other severe penalties for capital crimes—and that, among other organizations, both the American Academy of Pediatric Neuropsychology (AAPdN) and the American Psychological Association (APA) have called for raising the lower limit for the death penalty from 18 to 21 years (see also McCaffrey & Reynolds, 2021). More on this in a moment, after an overview of the key court decisions in Table 4.1.

The Scientific Argumentation

After this introductory overview, we will now deal with the topic in three steps: First, we will look at the scientific argumentation from today's perspective as to why juveniles, adolescents and/or young adults should be less criminally responsible—and thus would deserve lesser sentences. In the next step, we will work out the argumentation from the court decisions, particularly with regard to the development of young offenders. I will then evaluate the arguments from my point of view.

Not only do we have the advantage of having already dealt with important basics of these questions in Chapters 2 and 3, but we can also look back on 20 years of discussion. For the above-mentioned court rulings, various scientific and medical organizations have acted as *amicus curiae* (Lat. friend of the court) and submitted recommendations. Instead of having to delve into these documents—some of which are now decades old—we can focus on the current statements of the AAPdN (Mucci, 2021) and APA (Haney et al., 2022).

The AAPdN resolution fits on one page and can therefore be dealt with briefly. First, the psychologists summarize that *Roper* was based on three findings: Firstly, adolescents are not yet fully grown up and have an underdeveloped sense of responsibility; secondly, they are more susceptible to negative influences, such as peer pressure; and thirdly, their character is not yet as stable as that of adults. The Association derives its demand from this:

> The AAPdN believes the primary reason these findings are true and accurate is the level of maturity (or immaturity) of the brain at this age. However, there is no bright line regarding brain development nor is there neuroscience to indicate the brains of 18-year-olds differ in any significant way from those of 17-year-olds. An examination of the research on brain development indicates ongoing maturation of the brain through at least age 20. Thus, it is the opinion of the AAPdN that the same prohibitions applied to application of the death penalty to persons aged 17 should apply to persons ages 18 through 20 years and for the same scientific reasons. (Mucci, 2021, p. 88)

In the same year, the APA set up a working group to address the issue. Among its members were familiar names such as Arielle R. Baskin-Sommers from Yale University, B. J. Casey, also from Yale, Elizabeth E. Cauffman from the University of California at Irvine and Leah H. Somerville from Harvard University. Their proposal was made available to both APA members and the general public in 2022 and subsequently revised. On August 3, 2022, the APA Council voted 161:7 in favor of the final version. As in the AAPdN resolution, it states that there is no clear neuroscientific boundary in brain development between 17-year-olds on the one hand and 18- to 20-year-olds on the other. The psychologists and neuroscientists explained in more detail:

> [N]euroscientific research demonstrates brain development at age 17 has not become static and there is significant, ongoing brain development in the 'late adolescent class'. While some research on continued neurobiological development after 17 was published prior to the Roper decision, the question of whether members of the late adolescent class (ages 18 to 20) should be eligible for death as a penalty was not before [the Supreme Court] at the time of the Roper decision and thus was not considered. [...] [I]t is clear the brains of 18- to 20-year-olds are continuing to develop in key brain systems related to higher-order executive functions and self-control, such as planning ahead, weighing consequences of behavior, and emotional regulation. Their brain development cannot be distinguished reliably from that of 17-year-olds with regard to these key brain systems. (Haney et al., 2022, pp. A1–A2)

In line with our discussion in Chapter 2, numerous other psychological processes were mentioned in which development is not yet complete in this age group. Particularly relevant brain regions for this are the prefrontal cortex and the areas associated with it, for which there is "significant development [...] that continues beyond the age of 20" (ibid., p. A2). In total, the approximately three-page text of the "APA Resolution on the Imposition of Death as a Penalty for Persons Aged 18 Through 20" contains the prefix "neuro" ten times and the word "brain" 18 times.

Another interesting line of argumentation in the resolution points to numerous examples of legislation that particularly protects or restricts young adults up to the age of 20, for example with regard to the purchase of psychoactive substances or weapons and the holding of certain political offices. According to Alex Meggitt from Lewis & Clark Law School in Portland, there are even "over 3000 laws across the USA that limit a person's privileges or abilities based on not achieving the age of 21" (Meggitt, 2021, p. 74). We recall Robert Epstein's argument from Chapter 2 that adolescent problem behavior is also a consequence of the increasing prohibitions for this age group (Epstein, 2007). While Epstein used this as an argument against the concept of adolescence, the APA working group used it to justify the expansion of this life stage (Haney et al., 2022).

This unequal legal treatment could actually be significant for the Supreme Court. For *Roper*, the age limits in the US states for voting, serving on a jury and marrying without parental consent were compared in detail. These were 18 years of age almost everywhere. This was then also taken as the threshold for sufficient criminal responsibility for the imposition of the most severe sentence in the Court's decision. In this context, it is worth recalling the increase in criminal behavior in adolescents and young adults, which was previously discussed: In a review article, Laurence Steinberg compared the corresponding curves for 1990, 2000 and 2010. While, according to figures from the FBI, compared to 1990, the number of arrests for many types of crime fell by as much as half, the sharp spike toward the end of the teenage years remained (Steinberg, 2013). This suggests that we are dealing with both an individual developmental phenomenon and a social trend.

There are now extensive overviews of the role of the brain in the psychological development of adolescents, which have also dealt intensively with the normative significance of these differences (Casey et al., 2022; Haney et al., 2022). It seems to have become more natural for researchers to actively participate in socio-political debates. For example, Haney and colleagues pointed out that a disproportionately large share of the harshest penalties—especially the death penalty—in the USA is imposed on non-white sections of the population (Haney et al., 2022). Raising the age limit for full criminal responsibility to 21 could also reduce racial disadvantage, the authors argued. These researchers also discussed that the most severely convicted offenders were likely to have a particularly large gap in neural, cognitive and emotional development, as these factors depend, for example, on experienced trauma and other psychosocial problems (ibid.).

Casey and colleagues derived a further argument against the most severe penalties from the sharp rise in criminal behavior in adolescence and young adulthood: Since personality is not yet stabilized and crime decreases with age anyway, a life or even death sentence would affect many people who would not commit such serious crimes again in the future. The majority of these offenders would be deprived of the chance of a normal future in a disproportionately serious way, even though they did not pose a great danger to the general public (Casey et al., 2022). We have already said a great deal about psychological and neuronal development. The researchers' conclusion summarized the state of scientific knowledge in a few sentences:

Overall, the literature on the development of psychological abilities reveals two key findings. The first is that adolescents and young adults as a group show immature psychological abilities relative to adults, which justifies special treatment and protection of youth. The second is that there is no one age at which an individual reaches maturity in all psychological capacities The development of cognitive, emotional, and social psychological abilities mature at different ages and this development can extend beyond 18 years. As such, an adolescent may have the capacity to make rational decisions in one context but lack the ability to engage in mature decision-making in another. (ibid., pp. 328–329)

And they wrote specifically about the brain:

Thus, the idea that there is a single age when the brain is mature or no longer exhibits plasticity conflicts with neuroscientific evidence of continued changes. Moreover, there is tremendous variability in the age at which changes are observed in the brain [...]. Regardless of this variability, there are reliable brain changes that occur beyond age 18 that are relevant to criminal behavior and involve brain circuitry implicated in decision-making (e.g., prefrontal cortex). (ibid., p. 329)

Let us bear in mind that these researchers reject a strict age limit of 18 years—but also cannot propose a concrete upper threshold beyond which people are fully developed. Let us now look at the legal reasoning in the above-mentioned court rulings that are relevant to these questions.

The Judicial Reasoning

When introducing the scientific resolutions in the last section, we already briefly referred to the three aspects of adolescent development discussed in *Roper v. Simmons* (2005): firstly, their immaturity and underdeveloped sense of responsibility; secondly, their greater susceptibility to negative influences and peer pressure; and thirdly, their incomplete identity development. The research findings of the developmental psychologists Jeffrey J. Arnett and Laurence Steinberg, with whom we have been familiar since Chapter 2, played an important role here. However, the court decision bracketed this into the earlier case law and into what "any parent knows and as the scientific and sociological studies respondent and his *amici* cite tend to confirm" (ibid., p. 15).

What is important for us at this point are the *normative* conclusions. According to the Court, it is legally and morally problematic to classify young people as the worst offenders because of the differences mentioned above. Due to their greater vulnerability and less control, they would deserve more forgiveness than adults. The fact that their personal development is still progressing also would make it less likely that a young person's character is irretrievably "depraved." The limited culpability of juveniles called into question the reasons for justifying the death penalty: With regard to the aspect of retribution, it would not be proportionate to give the most severe punishment to the group with substantially reduced culpability; and with regard to deterrence, juveniles lacked precisely the mature ability to deliberate on the consequences of their actions.

However, according to the Supreme Court's decision, this is in contrast to the brutality of the crimes committed by some juvenile offenders. The scope for juries and courts to take into account individual factors of the offender or the offense had to be considered, too. But: "The differences between juvenile and adult offenders are too marked and well understood to risk allowing a youthful person to receive the death penalty despite insufficient culpability" (ibid., p. 19). A firm demarcation would be difficult, also from a psychological point of view, but necessary. The aforementioned characteristics of young people did not suddenly disappear when they turn 18. However: "The age of 18 is the point where society draws the line for many purposes between childhood and adulthood. It is, we conclude, the age at which the line for death eligibility ought to rest" (ibid., p. 20).

In the decision *Graham v. Florida* (2010), life imprisonment without the prospect of parole for minors was only found to be constitutional for homicide offenses. The reasoning is analogous to that in *Roper* and was extended to the second most severe penalty after the death penalty. What these two punishments have in common is that those convicted necessarily die in captivity. However, the opinion for *Graham* referred more explicitly to "developments in psychology and brain science," which "continue to show fundamental differences between juvenile and adult minds" (ibid., p. 17). This applied, for example, to areas of the brain associated with behavioral control. In addition to retribution and deterrence, the Court mentioned two further justifications of punishment, namely incapacitation (to commit further crimes) and rehabilitation.

Shortly thereafter, in *Miller v. Alabama* (2012), the constitutional prohibition was extended to homicide offenses if they *automatically* lead to life without the prospect of parole. In a footnote, the court stated, with reference to an APA opinion: "The evidence presented to us in these cases indicates that the science and social science supporting *Roper's* and *Graham's* conclusions have come to be even stronger" (ibid., p. 472n5). In particular, brain regions of adolescents in connection with functions such as impulse control, planning ahead and risk avoidance were not yet fully developed. Recent studies had also shown that contact with peers who display deviant behavior lead to an increase in such behavior and delinquency.

Although the Court did not generally rule out the severe punishment for minors, it called for individual consideration of their development, particularly with regard to the correctability of their character. As it is held in the decision: "Although we do not foreclose a sentencer's ability to make that judgment in homicide cases, we require it to take into account how children are different, and how those differences counsel against irrevocably sentencing them to a lifetime in prison" (ibid., p. 480). However, the required consideration would not be compatible with an *inevitable* sentencing for life without parole. Therefore, this form of punishment would also be "cruel and unusual"—and thus unconstitutional.

The decision, *Montgomery v. Louisiana* (2016), dealt more with the formal legal question of whether the conclusion of *Miller* must be applied retroactively. Although the Court discussed the previous rulings, it no longer specifically addressed scientific findings. Terms such as "science," "brain," "neuro" or "psychology" do not appear in the entire opinion and even "adolescent" only once. The question of the maximum permissible sentence for offenders who were minors at the time of the offense now came down to the aspect of whether someone was incorrigibly a criminal or not:

Before *Miller*, every juvenile convicted of a homicide offense could be sentenced to life without parole. After *Miller*, it will be the rare juvenile offender who can receive that same sentence. The only difference between *Roper* and *Graham*, on the one hand, and *Miller*, on the other hand, is that *Miller* drew a line between children whose crimes reflect transient immaturity and those rare children whose crimes reflect irreparable corruption. The fact that life without parole could be a proportionate sentence for the latter kind of juvenile offender does not mean that all other children imprisoned under a disproportionate sentence have not suffered the deprivation of a substantive right. (ibid., p. 18)

The applicant in this case, Henry Montgomery, had shot a police officer in 1963 at the age of 17. He received the death penalty for murder in the first trial in 1964. However, this verdict was overturned due to procedural errors and was followed by another conviction for murder in 1969, but this time with an automatic life sentence without parole. The Supreme Court's decision allowed him to apply for parole—and he was released in 2021 after almost 58 years in prison. The individual fates behind the long prison sentences have also been addressed in recent research (Casey et al., 2022; Haney et al., 2022).

In the (so far) last case in this series, an at the time of the crime 15-year-old murderer convicted in 2004 wanted to enforce that there should be a separate investigation into his incorrigibility in the criminal sense. In *Jones v. Mississippi* (2021), however, the Supreme Court rejected this: According to *Miller* and *Montgomery* the courts must take into account the age of the young offenders. However, a specific procedure and, in particular, a separate investigation were not prescribed. Because there were other options in this case besides life without parole, this punishment was not *inevitable* and thus also not unconstitutional, the Court argued. If the rules of a US state allow it, the consideration of this punishment therefore remains within the competence of the jury and the courts.

In this decision, too, as in *Montgomery,* there is no reference to scientific studies. But at that time the normatively relevant difference between minors and adults was not at issue. However, under Donald Trump's presidency from 2017 to 2021, the majority on the US Supreme Court had changed and all six judges who were considered to belong to the conservative camp supported the decision with three dissenting votes.

Evaluation

We can summarize that the consideration of the development of minors for criminal responsibility in capital cases is now established law in the USA. Scientific studies have repeatedly been used to justify this. *Graham* and *Miller* also explicitly cited findings from brain research. However, this was not uncontroversial among the Supreme Court judges. Already with *Roper*, for example, Justice Antonin G. Scalia (1936–2016) pointed out an apparent inconsistency in his dissenting opinion: The APA introduced findings that argued against the responsibility of 16- and 17-year-olds, but for a decision in favor of the right for abortion for minors, *Hodgson v. Minnesota* in 1990, it would have argued the opposite, citing studies that even 14- and 15-year-olds could think about moral issues and laws in a similar way to adults.

Laurence Steinberg and colleagues subsequently addressed the accusation of selectively choosing scientific studies for political-liberal reasons. They also conceded that scientific authority on political issues depends crucially on the credibility of their findings (Grisso & Steinberg, 2005). Based on data we already discussed in Chapter 2, developmental psychologists introduced a distinction here: Even if adolescents reached adult levels faster in the area of cognitive abilities, their social-emotional development, impulsivity and influenceability by peer pressure take more time to develop (Steinberg et al., 2009). In addition, the decision for or against an abortion would usually involve more time for consideration. The researchers here actually introduced a dual process model (i.e., distinguishing between fast and slow modes of decision making) and held that adolescents and young adults should be treated differently by legal institutions because of their impaired social-emotional capacities (Scott et al., 2016).

In another review, which dealt intensively with the decisions of the Supreme Court up to 2012, Steinberg specified this further: While the cognitive control system developed linearly in adolescents, the system for processing incentives showed a more rapid development in the teenage years, with a peak at around 15–18 years of age (Steinberg, 2013). This resulted in an increased susceptibility to risky behavior in adolescence. A similar pattern could be seen in impulse control versus sensation seeking.

Based on these and other studies, one *can* justify the separation of the different age groups scientifically. However, whether a certain normative distinction *must follow* from this is a complicated question. For example, Fig. 2.4 showed a small 0.25 and 0.15 point difference, respectively, of 16- to 17- and 18- to 21-year-olds compared to people aged 22–25 in the social-emotional domain. How big such a difference must be to justify a categorical normative distinction is hard to tell—and a question that cannot be answered by science alone. As we have seen, the related Supreme Court decisions were often only taken by a narrow majority and possibly reflected the justices' political alignment (Table 4.1).

Before addressing fundamental problems in the transfer from science to law, let us discuss the specific contribution of neuroscience. As we saw in Chapter 3, phrenologists had a great interest in being taken seriously in court already in the early nineteenth century. After *Roper*, there were sometimes fanciful speculations, especially in the neurolaw literature, about the extent to which the Supreme Court was influenced by brain research. Here are just two examples: "Scientific evidence about the developing brain formed the basis for Eighth Amendment protections against executing juveniles (Roper v. Simmons, 2005) […]" (Khalid et al., 2024, p. 66) and:

> While not quoted directly, briefs submitted by the American Medical Association and the American Psychological Association on behalf of the defendant, explaining the current state of research on adolescent brain immaturity, had clearly informed the judges' verdict. (Harman, 2013, p. 457)

Anyone who remembers the quotes in the previous section will probably regard such statements as speculative and exaggerated. Reference is sometimes made to the oral hearing of *Roper v. Simmons* (2005), in which Seth P. Waxman, an attorney for Christopher Simmons, stated: "[A]nd I'm not just talking about social science here, but the important neurobiological science that has now shown that these adolescents are – their character is not hard-wired" (ibid., p. H40). This was a specific answer to Justice Breyer's question: "Now, I thought that the – the scientific evidence simply corroborated something that every parent already knows, and if it's more than that, I would like to know what more" (ibid.). To deduce from this that the neurosciences were decisive or even relevant for the judgment is, in my opinion, a risky move. Stephen J. Morse criticized such statements already in 2006 as "brain overclaim syndrome" (Morse, 2006).

I think Steinberg summarized it aptly: "[N]euroscience may have played a part in persuading the justices that the psychological differences between adolescents and adults as described in *Roper* were genuine and indisputable" (Steinberg, 2013, p. 517). A systematic comparison of such statements in the scientific literature, as I have done for the presentation of Phineas Gage (Schleim, 2022), could be interesting. One should also consider the potential conflict of interest that researchers have on this issue. Attention from the highest courts is not only associated with prestige and thus possibly a better chance of obtaining scarce funding. Scientists can also earn high additional income as expert witnesses (e.g. Harman, 2013).

We should ask ourselves whether the Supreme Court could have made the decisions without the (neuro-) scientific studies. The answer would be "yes." After all, differences between adults and minors are also visible in everyday experience, which is not least what the established age limit of 18 is based on. As we have seen, the police statistics of various countries also portray young people as a special group. But given the narrow majorities in these important decisions, a reference to brain research might be particularly persuasive. We will return to this idea in Chapter 5.

According to my analysis, three fundamental problems remain when drawing conclusions from such scientific data on norms: (1) the individual variability within an age group, (2) the lack of a neuroscientifically based age limit and—related to this—(3) the lack of practical relevance. As I believe these problems apply to all three examples in this chapter, they will be addressed jointly in the summary.

4.3   The "Cannabis Brain" in Germany

We briefly mentioned in the last section that brain development was occasionally referred to when adapting laws on psychoactive substances in the USA (see Meggitt, 2021). In Germany—after a tough public and parliamentary debate on public health—a limited decriminalization of the cultivation and possession of psychoactive cannabis came into force on April 1, 2024 (Schleim, 2024b). One of the reasons for this initiative was that consumption had increased despite the ban and that illegal products may be more harmful and promote crime due to a lack of control. The plan of the Social Democratic, Green and Liberal government coalition to make possession from the age of 18 exempt from punishment was criticized above all by the medical profession. The German Society for Psychiatry and Psychotherapy, Psychosomatics and Neurology (DGPPN) gave the most detailed reasons for its reservations in its statement of November 2, 2023, in which it held:

> The age limit for access to cannabis is too low at 18 years, as brain development is generally not yet complete by the mid-20s. Due to consistent clinical findings on increased risk of psychosis and altered maturation of neurons with early cannabis use, e.g. in adolescence and young adulthood, cannabis should not be consumed before brain maturation is complete. [...] Brain maturation is not completed until the middle of the third decade of life, with large inter-individual differences [...]. Due to consistent clinical

findings on increased risk of psychosis and altered maturation of neurons and myelination with early cannabis use, e.g. in adolescence and young adulthood in clinical and experimental studies, cannabis should not be used before the brain has completed maturation [...]. **From a psychiatric and neurobiological point of view and the current state of knowledge, the age limit for access should therefore not be below 21 years.** (DGPPN, 2023, p. 1, 3; emphasis in original)

Nevertheless, the law that has been passed makes possession of the substance largely exempt from punishment from the age of 18. From the age of 18–20 inclusive, however, only smaller quantities may be dispensed in special cannabis associations, and then only products with a lower THC concentration (the major psychoactive ingredient in cannabis products). According to official figures, the drug was most frequently consumed by 18- to 24-year-olds in the past, despite the ban. The regulations on consumption in road traffic are currently still being worked on, but will probably provide for a complete ban under the age of 21. Despite the lower age limit in the law, the German Federal Ministry of Health states: "The consumption of cannabis poses health risks, especially for children and adolescents, as [...] the human brain is particularly vulnerable until maturity at the age of 25."[11]

What is interesting for us is not only that brain maturity was set at 25 years. The DGPPN statement also mentioned a specific mechanism for this, namely myelination. As we saw in Chapter 2, this process might actually occur more frequently from the mid-20s onwards. The brain is never completely finished, but remains a plastic organ until the end of life. In the following, we look at the studies on brain development cited by the DGPPN and link this to alleged brain changes in connection with cannabis use.

Cannabis, Brain Development and Psychosis

The DGPPN cited three studies as evidence of the "generally not yet completed by the mid-20s" brain development: The first publication dealt with the myelination of mammalian cells, not specifically the human brain;

---

[11] The German Federal Ministry of Health, which lead this initiative, provides basic and updated information on the law at: https://www.bundesgesundheitsministerium.de/themen/cannabis/faq-cannabisgesetz.

accordingly, it does not mention an age limit of 25 years (Baumann & Pham-Dinh, 2001). But with reference to a study from the 1960s, it was mentioned that myelination continues in some areas of the brain until the age of 20. This publication, however, reported a completed development of the nervous system before 20, with the exception of the association cortex, which could, however, develop beyond the age of 30 (Yakovlev & Lecours, 1967).

The second publication cited by the DGPPN did not examine brain development, but cognitive impairments due to cannabis use; such were reported, but were no longer statistically significant after 72 hours of abstinence (Scott et al., 2018). However, these researchers referred to two studies on brain development, the first of which examined test subjects up to the age of 21 and the second up to the age of 22 (Giedd et al., 1999; Satterthwaite et al., 2013). An age limit of 25 years cannot therefore be demonstrated with such studies for fundamental reasons and we have already discussed more recent and more detailed research in Chapter 2.

The third and final citation from the DGPPN's resolution examined the brain development of eight- to 21-year-olds in connection with poverty and stressful life events (Gur et al., 2019). Again, this does not allow any firm conclusions about the mid-20s. Yet, this study reported the generally interesting finding for our book that a difficult childhood, to summarize briefly, accelerates brain maturation. The researchers explained this in terms of evolutionary biology, stating that adverse circumstances required earlier adult reactions. Due to various neurobiological markers, the brains of minors (8–17 years old) were then more like those of adults (18 years and older). This fits in with our discussion of parentification in Chapter 2. But the correlations were sometimes complex: For example, low affluence was associated with a *lower* density of gray matter, but stressful experiences with a *higher* density (ibid.). It should not be inferred from this that one effect can be "treated" by the other. In any case, such findings are initially only correlations that do not provide conclusive causal evidence in themselves.

This applies equally to the frequently reported link between cannabis use and psychosis or even a diagnosis of schizophrenia. This is particularly assumed for heavy and frequent substance use (e.g. Hasan et al., 2020). However, observational studies do not allow conclusions to be drawn about causal relationships for fundamental reasons and it is difficult to conduct experiments on this for ethical and methodological reasons. We also know that people at risk of psychosis use cannabis more frequently, partly to alleviate symptoms and partly to compensate for the side effects of antipsychotic medication (e.g. Carrión et al., 2023). Self-medication or coping is one of many reasons for substance use (Schleim, 2023). But epidemiological data from Denmark, for example, also show that the group of schizophrenia patients is much larger than that of heavy cannabis users with a diagnosis of schizophrenia. Actually, there were only 6050 cases (corresponding to 0.09 percent of the almost 7 million investigated people) in that latter group for the period of 1972 to 2021 (Fig. 4.1; Hjorthøj et al., 2023).

One problem in the drug policy debate is that the prohibitionist status quo is often compared with the practically unrealistic state of complete abstinence. From the perspective of instrumental substance use, it would have to be clarified which other means people then use to achieve their aims—and whether these are more or less harmful (Müller, 2020; Schleim, 2023, 2024b). Instead, different realistic drug policy options should be compared. In any case, a recent epidemiological study of the health data of over 63 million Americans found no statistically significant

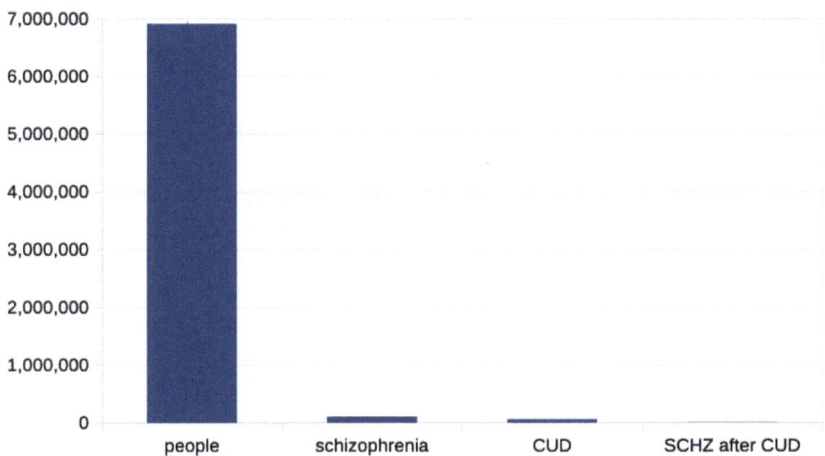

Fig. 4.1 According to Danish health data for the years 1972–2021, a diagnosis of schizophrenia is much more common in the overall population with an assumed prevalence of 1.5%, but relatively rare after heavy cannabis use (cannabis use disorder = CUD). Causal conclusions should be interpreted with caution (*Source* Hjorthøj et al. [2023])

increase in medical treatment for psychosis following cannabis legalization (Elser et al., 2023).

Instead of discussing drug policy, which would take us away from the topic of the book, we will conclude with the question of whether there is such a thing as a "cannabis brain," that is, the extent to which long-term use of the substance is reflected in brain changes. This question was investigated, among others, by the psychiatrists J. Cobb Scott and Ruben C. Gur from the University of Pennsylvania and colleagues cited before and by the DGPPN. For a longitudinal study of almost 800 adolescents and young adults aged 14–22, 147 of whom were cannabis users, they carried out structural MRI brain scans. Their conclusion:

> There were no significant differences by cannabis group in global or regional brain volumes, cortical thickness, or gray matter density, and no significant group by age interactions were found. Follow-up analyses indicated that values of structural neuroimaging measures by cannabis group were similar across regions, and any differences among groups were likely of a small magnitude. In sum, structural brain metrics were largely similar among adolescent and young adult cannabis users and non-users. Our data converge with prior large-scale studies suggesting small or limited associations between cannabis use and structural brain measures in youth. (Scott et al., 2019, p. 1362)

Now these people were relatively young when the brain images were taken. The study of over 1000 people from Dunedin, New Zealand, which has been running since 1972, is therefore informative with regard to possible brain changes after a longer period of time. The long-term cannabis users had a slightly lower IQ in their 40s than estimates from childhood and puberty would suggest. However, these differences could not be explained by brain changes. The researchers only found a slightly lower volume in the hippocampi, which are often associated with memory, when they specifically analyzed this brain region (Meier et al., 2022). But this rather modest finding could not be confirmed in another study by the same research group from the same year when the brain was analyzed as a whole. It is then more difficult for small effects to reach the necessary statistical threshold because stronger corrections have to be made to avoid false-positive findings. The brain changes they then found could be better explained by the substance users' alcohol and tobacco consumption (Knodt et al., 2022).

Even if further studies are needed, we can state the following here: As with the Dutch "criminal neurolaw," the cited neuroscientific studies on brain development do *not* support the cited age limits—this time: 25 years. Again, these were actually misquotations, which I would consider a serious error in a bachelor thesis. In my opinion, speaking of "consistent clinical findings," as the DGPPN did, is thus wrong and misleading.

The discussion of the risk of psychosis and the alleged "cannabis brain" shows once again how difficult it is to transfer study results to practice (Schleim & Roiser, 2009). In any case, severe side effects from cannabis use seem to occur rather rarely, depend on the individual risk of psychosis and are not directly reflected in brain development. Incidentally, researchers have a tendency to quickly portray possible changes associated with cannabis negatively as "abnormality," "dysfunction" or "disorder," even if they cannot link the brain findings to psychological problems (e.g. Knodt et al., 2022). By contrast, in a recent study on the changes in brain function caused by frequent consumption of sweet or fatty meals, researchers spoke of an "enhanced response" of the reward system in the brain (Thanarajah et al., 2023). We should be careful about projecting our personal norms onto brain data.

4.4   Summary

After discussing some of the more fundamental questions of neurolaw, brain research and criminal responsibility in Chapter 3, this chapter is devoted in particular to normative questions relating to brain development. As I pointed out, there was a neuroscience-backed legislative initiative in the Netherlands as early as 2012, to my knowledge the first of its kind in the world, to extend juvenile criminal law to up to 22 years in individual cases. According to the findings discussed in Chapter 2, one *can* argue that way—but as shown here, the specifically cited publications from brain research did *not* support the legislator's respective claims. We found the same pattern with the German "cannabis brain." Here too, as with some US laws, the argument was simply that brain development continues up into the 20s.

The examples discussed in this chapter all had in common to postpone a legally relevant distinction to an older age based on findings about the ongoing brain development of adolescents and young adults. According to my analysis, such efforts run into three fundamental problems: firstly, individual variability within an age group; secondly, the lack of a neuroscientifically based age limit and, related to this, thirdly, the lack of practical relevance. In more detail:

Firstly, the data on psychological and neurobiological development reflect mean differences between different age groups. However, a closer look reveals large individual differences within a group. We already discussed this briefly in Chapter 2 (Fig. 2.6), but a more recent publication to determine the cognitive maturity of adolescents also shows this. This combined an IQ test, tasks on cognitive control, emotional information processing and risk-taking behavior (Fig. 4.2; El Damaty et al., 2022).

Casey and colleagues, who argued in favor of raising the minimum age for the death penalty to 21, also acknowledged this: The variability "within a single age was as large as the variance between ages.

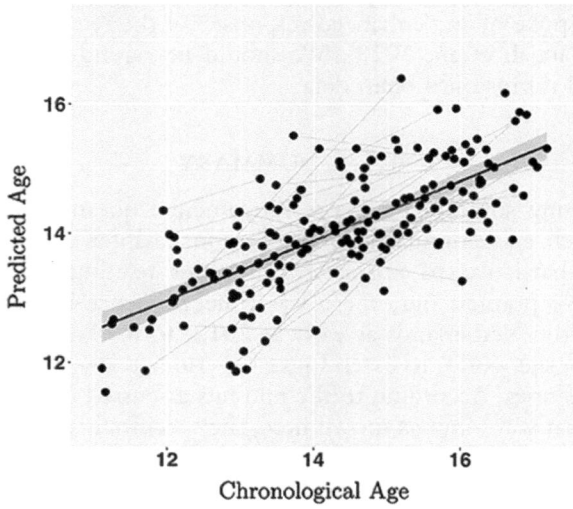

Fig. 4.2 Combination of various emotional and cognitive tests shows a strong correlation between age and "neurocognitive age." The data show, for example, that individual 12-year-olds can exhibit the maturity of 14- to 16-year-olds and vice versa (*Source* El Damaty et al. [2022]. License: CC BY 4.0 [http://creativecommons.org/licenses/by/4.0/])

Distinguishing the capacity of a 17-year-old from an 18-, 19-, 20-, or 21-year-old would be impossible for a single individual or even group of individuals" (Casey et al., 2022, p. 327). And other researchers stated: "More importantly, adolescent brain immaturities should not be interpreted as evidence to exonerate all responsibility, especially considering that not all adolescents commit crimes" (Luna & Wright, 2016, p. 108). Accordingly, Supreme Court Justices Scalia and Thomas criticized in their minority decisions for *Roper*, *Graham* and *Miller* that the restrictions on maximum sentences refer categorically to adolescents as a group, whereas there are also responsible minors. In *Graham*, Justice Thomas even cited Epstein's *The Case Against Adolescence*, that we discussed in Chapter 2. In *Roper v. Simmons* (2005), Justice Scalia argued:

> Even putting aside questions of methodology, the studies cited by the Court offer scant support for a categorical prohibition of the death penalty for murderers under 18. At most, these studies conclude that, on average, or in most cases, persons under 18 are unable to take moral responsibility for their actions. Not one of the cited studies opines that all individuals under 18 are unable to appreciate the nature of their crimes. (ibid., p. D12)

It is fitting that Christopher Simmons' crime was not an impulsive, emotional crime, but planned in advance; and that his 16-year-old friend stopped his involvement in time. In the case of Evan Miller, who was only 14 years old at the time of the crime, the robbery with severe bodily harm was committed out of the situation and under the influence of drugs. However, the perpetrator returned later and set fire to destroy evidence. His victim, who was seriously injured at the time, also died as a result of smoke inhalation (see e.g. Hirstein et al., 2018). If such individual distinctions are possible, why should juries and courts not also be able to deliberate on the appropriate punishment in individual cases? The fact that different points of view on this are legally possible is shown by the narrow majorities of the decisions (Table 4.1).

Secondly, the arguments from a scientific point of view always amounted to a shift of the age limit backwards, without being able to name a concrete age for the ideal demarcation. Steinberg and colleagues themselves stated: "The notion that a single line can be drawn between adolescence and adulthood for different purposes under the law is at odds with developmental science" (Steinberg et al., 2009, p. 583). The fact that the boundary must lie "somewhere higher" is partly due to the individual variability just mentioned. In addition, the complexity of quantifying psychological and neurobiological development in concrete terms is also expressed in the problem. As we saw in Chapter 2 and again here, science allows different boundaries to be drawn. But instead of just criticizing the status quo, a positive proposal would be helpful.

Nevertheless, if society's views on when adulthood begins change, this is likely to be reflected in new legal boundaries in the long term. We saw earlier that this process has been going on for some 200 years. But there are two potential errors associated with each determination: On the one hand, responsible adolescents are regarded as immature, while on the other hand, irresponsible adolescents are regarded as mature. As in the methodology of statistics, it is a question of weighing up which error one wants to minimize (namely false-positive or false-negative results). On the other hand, if one wanted to fundamentally individualize the decisions, then all the examples discussed in this chapter would fall away, as they drew categorical distinctions. This leads us to the last point.

Thirdly, the practical added value by including neuroscience is still missing from the examples mentioned here. It is one thing to point to the continuing development of the brain as a justification. In these cases, however, it seems to amount to a more socio-political demand, "We think it's too young!" which is given more weight by referring to brain research. It is quite another matter to determine the development of a person in an individual case, for example in a specific criminal trial. Based on the findings presented in Chapter 2, this would be possible in principle. Simply determining age neurobiologically would not have any relevant additional benefit, as this variable is generally undisputed. With regard to psychological maturity, brain research methods would first have to show that they enable better decisions to be made than the existing methods. Such measurements would be subject to a certain degree of inaccuracy and it is likely that experts would disagree about their validity in court.

Individual circumstances are taken particularly seriously in criminal proceedings with their impending consequences—and the possibilities for individual assessment are therefore particularly suitable. In the long term, new practices could be established in this way. We will return to the benefit aspect in the final chapter, but only then, in my view, would the existence of a useful, coherent and meaningful new practice be given (Schleim, 2024c). Under the present circumstances, though, also considering the current conservative majority on the Supreme Court, a further increase in the minimum age for the most severe sentences in the USA seems unrealistic to me. Let us end this chapter with a quote from Steinberg:

> However, I believe that in discussions of where we should draw legal boundaries between adolescents and adults, neuroscience should continue to have a supporting role, and behavioral science should continue to carry the weight of the argument. Ultimately, the law is concerned with how we behave and not with how our brains function. (Steinberg, 2013, p. 517)

REFERENCES

Adleman, N. E., Menon, V., Blasey, C. M., White, C. D., Warsofsky, I. S., Glover, G. H., & Reiss, A. L. (2002). A developmental fMRI study of the Stroop color-word task. *NeuroImage, 16*(1), 61–75.

Baumann, N., & Pham-Dinh, D. (2001). Biology of oligodendrocyte and myelin in the mammalian central nervous system. *Physiological Reviews, 81*(2), 871–927.

Carrión, R. E., Auther, A. M., McLaughlin, D., Adelsheim, S., Burton, C. Z., Carter, C. S., et al. (2023). Recreational cannabis use over time in individuals at clinical high risk for psychosis: Lack of associations with symptom, neurocognitive, functioning, and treatment patterns. *Psychiatry Research, 328*, 115420.

Casey, B. J., Getz, S., & Galvan, A. (2008). The adolescent brain. *Developmental Review, 28*(1), 62–77.

Casey, B. J., Simmons, C., Somerville, L. H., & Baskin-Sommers, A. (2022). Making the sentencing case: Psychological and neuroscientific evidence for expanding the age of youthful offenders. *Annual Review of Criminology, 5*, 321–343.

Casey, B. J., Tottenham, N., Liston, C., & Durston, S. (2005). Imaging the developing brain: What have we learned about cognitive development? *Trends in Cognitive Sciences, 9*(3), 104–110.

Chandler, J. A., Harrel, N., & Potkonjak, T. (2019). Neurolaw today—A systematic review of the recent law and neuroscience literature. *International Journal of Law and Psychiatry, 65*, 101341.

Cohen, A. O., & Casey, B. J. (2014). Rewiring juvenile justice: The intersection of developmental neuroscience and legal policy. *Trends in Cognitive Sciences, 18*(2), 63–65.

DGPPN [Deutsche Gesellschaft für Psychiatrie und Psychotherapie, Psychosomatik und Nervenheilkunde]. (2023). *Stellungnahme der DGPPN zum Entwurf eines Gesetzes zum kontrollierten Umgang mit Cannabis.* DGPPN.

Dreyfuss, M., Caudle, K., Drysdale, A. T., Johnston, N. E., Cohen, A. O., Somerville, L. H., et al. (2014). Teens impulsively react rather than retreat from threat. *Developmental Neuroscience, 36*(3–4), 220–227.

El Damaty, S., Darcey, V. L., McQuaid, G. A., Picci, G., Stoianova, M., Mucciarone, V., et al. (2022). Introducing an adolescent cognitive maturity index. *Frontiers in Psychology, 13,* 1017317.

Elser, H., Humphreys, K., Kiang, M. V., Mehta, S., Yoon, J. H., Faustman, W. O., & Matthay, E. C. (2023). State cannabis legalization and psychosis-related health care utilization. *JAMA Network Open, 6*(1), e2252689.

Epstein, R. (2007). *The Case Against Adolescence: Rediscovering the Adult in Every Teen.* Sanger, CA: Quill Driver Books.

Flanders, C. (2023). Cruel and unusual punishment. In M. C. Altman (Ed.), *The Palgrave handbook on the philosophy of punishment* (pp. 771–788). Palgrave Macmillan.

Giedd, J. N., Blumenthal, J., Jeffries, N. O., Castellanos, F. X., Liu, H., Zijdenbos, A., et al. (1999). Brain development during childhood and adolescence: A longitudinal MRI study. *Nature Neuroscience, 2*(10), 861–863.

Graham v. Florida, Opinion of the Court (2010). 560 U.S. 48.

Grisso, T., & Steinberg, L. (2005). Between a rock and a soft place: Developmental research and the child advocacy process. *Journal of Clinical Child and Adolescent Psychology, 34*(4), 619–627.

Gur, R. E., Moore, T. M., Rosen, A. F., Barzilay, R., Roalf, D. R., Calkins, M. E., et al. (2019). Burden of environmental adversity associated with psychopathology, maturation, and brain behavior parameters in youths. *JAMA Psychiatry, 76*(9), 966–975.

Haney, C., Baumgartner, F. R., & Steele, K. (2022). Roper and race: The nature and effects of death penalty exclusions for juveniles and the "late adolescent class" (with appendix). *Journal of Pediatric Neuropsychology, 8*(4), 168–177.

Harman, O. (2013). Unformed minds: Juveniles, neuroscience, and the law. *Studies in History and Philosophy of Science Part C: Studies in History and Philosophy of Biological and Biomedical Sciences, 44*(3), 455–459.

Hasan, A., von Keller, R., Friemel, C. M., Hall, W., Schneider, M., Koethe, D., et al. (2020). Cannabis use and psychosis: A review of reviews. *European Archives of Psychiatry and Clinical Neuroscience, 270,* 403–412.

Hirstein, W., Sifferd, K. L., & Fagan, T. K. (2018). *Responsible brains: Neuroscience, law, and human culpability.* MIT Press.

Hjorthøj, C., Compton, W., Starzer, M., Nordholm, D., Einstein, E., Erlangsen, A., et al. (2023). Association between cannabis use disorder and schizophrenia stronger in young males than in females. *Psychological Medicine, 53*(15), 7322–7328.

Hopman, M., & de Vocht, D. (2019). Criminal law for young adults in the Netherlands: The Law and the practice from the sociology of childhood perspective. In J. Hage, A. Walterman, & D. Roef (Eds.), *Law, science, rationality* (pp. 265–292). Boom.

Jones v. Mississippi, Opinion of the Court (2021). 593 U.S. ___.

Khalid, Z., Lee, R., & Wall, B. W. (2024). The use of neurobiological evidence in sentencing mitigation. *Behavioral Sciences & the Law, 42*, 65–78.

Knodt, A. R., Meier, M. H., Ambler, A., Gehred, M. Z., Harrington, H., Ireland, D., et al. (2022). Diminished structural brain integrity in long-term cannabis users reflects a history of polysubstance use. *Biological Psychiatry, 92*(11), 861–870.

Luna, B., & Wright, C. (2016). Adolescent brain development: Implications for the juvenile criminal justice system. In K. E. Heilbrun, D. E. DeMatteo, & N. E. Goldstein (Eds.), *APA handbook of psychology and juvenile justice* (pp. 91–116). American Psychological Association.

Matthews, S., Schiraldi, V., & Chester, L. (2018). Youth justice in Europe: Experience of Germany, the Netherlands, and Croatia in providing developmentally appropriate responses to emerging adults in the criminal justice system. *Justice Evaluation Journal, 1*(1), 59–81.

McCaffrey, R. J., & Reynolds, C. R. (2021). Neuroscience and death as a penalty for late adolescents. *Journal of Pediatric Neuropsychology, 7*, 3–8.

Meggitt, A. (2021). Trends in laws governing the behavior of late adolescents up to age 21 since Roper. *Journal of Pediatric Neuropsychology, 7*(1), 74–87.

Meier, M. H., Caspi, A. R., Knodt, A., Hall, W., Ambler, A., Harrington, H., et al. (2022). Long-term cannabis use and cognitive reserves and hippocampal volume in midlife. *American Journal of Psychiatry, 179*(5), 362–374.

Meltsner, M. (1973/2011). *Cruel and unusual: The Supreme Court and capital punishment*. Quid Pro Books.

Miller v. Alabama, Opinion of the Court (2012). 567 U.S. 460.

Moffitt, T. E. (2018). Male antisocial behaviour in adolescence and beyond. *Nature Human Behaviour, 2*(3), 177–186.

Morse, S. J. (2006). Brain overclaim syndrome and criminal responsibility: A diagnostic note. *Ohio St. J. Crim. l., 3*, 397–412.

Mucci, G. A. (2021). Resolution of the AMERICAN ACADEMY OF PEDIATRIC NEUROPSYCHOLOGY relating to the imposition of death as a penalty for persons ages 18 years through 20 years. *Journal of Pediatric Neuropsychology, 7*, 88.

Müller, C. P. (2020). Drug instrumentalization. *Behavioural Brain Research, 390*, 112672.

Paus, T., Collins, D. L., Evans, A. C., Leonard, G., Pike, B., & Zijdenbos, A. (2001). Maturation of white matter in the human brain: A review of magnetic resonance studies. *Brain Research Bulletin, 54*(3), 255–266.

Prop, L. J. C., Beerthuizen, M. G. J. C., & Van der Laan, A. M. (2021). *Adolescentenstrafrecht: Effecten van de toepassing van het jeugdstrafrecht bij jongvolwassenen op resocialisatie en recidive*. Wetenschappelijk Onderzoek- en Datacentrum.

Roper v. Simmons, Opinion of the Court (with [H]earing protocol and [D]issent Scalia) (2005). 543 U.S. 551.

RSJ [Raad voor Strafrechtstoepassing en Jeugdbescherming]. (2011). *Het jeugdstrafproces: toekomstbestendig*. RSJ. https://www.rsj.nl/binaries/rsj/doc umenten/rapporten/2011/03/14/advies-het-jeugdstrafproces-toekomstbest endig/Advies+Het+jeugdstrafproces+-+toekomstbestendig_RSJ_2011.03.14. pdf

Satterthwaite, T. D., Wolf, D. H., Erus, G., Ruparel, K., Elliott, M. A., Gennatas, E. D., et al. (2013). Functional maturation of the executive system during adolescence. *Journal of Neuroscience, 33*(41), 16249–16261.

Schleim, S. (2019). 'Neurorecht' in Nederland: De motivering van het nieuwe adolescentenstrafrecht vanuit een neurofilosofisch perspectief. *Algemeen Nederlands Tijdschrift voor Wijsbegeerte, 111*(3), 379–404.

Schleim, S. (2020). Real neurolaw in the Netherlands: The role of the developing brain in the new adolescent criminal law. *Frontiers in Psychology, 11*, 549375.

Schleim, S. (2022). Neuroscience education begins with good science: Communication about Phineas Gage (1823–1860), one of neurology's most-famous patients, in scientific articles. *Frontiers in Human Neuroscience, 16*, 734174.

Schleim, S. (2023). *Mental health and enhancement: Substance use and its social implications*. Palgrave Macmillan.

Schleim, S. (2024a). De bijzondere rol van de neurowetenschappen in het Nederlandse strafrecht. *Podium voor Bio-ethiek, 31*(1), 18–24.

Schleim, S. (2024b). *The German Cannabis protocols: Medicine, politics and science put to the test*. Stephan Schleim Philosophie und Psychologie.

Schleim, S. (2024c). *Science and free will: Neurophilosophical controversies and what it means to be human*. Springer.

Schleim, S., & Roiser, J. P. (2009). FMRI in translation: The challenges facing real-world applications. *Frontiers in Human Neuroscience, 3*, 845.

Scott, E. S., Bonnie, R. J., & Steinberg, L. (2016). Young adulthood as a transitional legal category: Science, social change, and justice policy. *Fordham Law Review, 85*, 641–666.

Scott, J. C., Rosen, A. F., Moore, T. M., Roalf, D. R., Satterthwaite, T. D., Calkins, M. E., et al. (2019). Cannabis use in youth is associated with limited alterations in brain structure. *Neuropsychopharmacology, 44*(8), 1362–1369.

Scott, J. C., Slomiak, S. T., Jones, J. D., Rosen, A. F., Moore, T. M., & Gur, R. C. (2018). Association of cannabis with cognitive functioning in adolescents and young adults: A systematic review and meta-analysis. *JAMA Psychiatry, 75*(6), 585–595.

Spanjaard, H. J. M., Filé, L. L., Noom, M. J., & Buysse, W. H. (2020). *Achterlopende ontwikkeling: Het begrip 'onvoltooide ontwikkeling' in de toepassing van het adolescentenstrafrecht.* Universiteit van Amsterdam.

Steiker, C. S., & Steiker, J. M. (2016). *Courting death: The Supreme Court and capital punishment.* Harvard University Press.

Steinberg, L. (2013). The influence of neuroscience on US Supreme Court decisions about adolescents' criminal culpability. *Nature Reviews Neuroscience, 14*(7), 513–518.

Steinberg, L., Cauffman, E., Woolard, J., Graham, S., & Banich, M. (2009). Are adolescents less mature than adults?: Minors' access to abortion, the juvenile death penalty, and the alleged APA "flip-flop." *American Psychologist, 64*(7), 583–594.

Thanarajah, S. E., DiFeliceantonio, A. G., Albus, K., Kuzmanovic, B., Rigoux, L., Iglesias, S., et al. (2023). Habitual daily intake of a sweet and fatty snack modulates reward processing in humans. *Cell Metabolism, 35*(4), 571–584.

Thompson, R. A., & Nelson, C. A. (2001). Developmental science and the media: Early brain development. *American Psychologist, 56*(1), 5.

Van der Laan, A. M., Beerthuizen, M. G., & Barendregt, C. S. (2021a). Juvenile sanctions for young adults in the Netherlands: A developmental perspective. *European Journal of Criminology, 18*(4), 526–546.

Van der Laan, A. M., Zeijlmans, K., Beerthuizen, M. G. J. C., & Prop, L. J. C. (2021). *Evaluatie van het adolescentenstrafrecht.* Wetenschappelijk Onderzoek- en Datacentrum.

Yakovlev P. I., & Lecours, A. R. (1967). The myelinogenic cycles of regional maturation of the brain. In A. Minkovski (Ed.), *Regional development of the brain in early life* (pp. 3–70). Blackwell.

CHAPTER 5

Brain and Behavior: A Pragmatic Approach

The law's criteria are virtually all behavioral – acts and mental states. This is especially true in criminal law or in any other legal context in which responsibility and competence are in question. (law and psychology professor Stephen J. Morse from the University of Pennsylvania; Morse, 2023, p. 235)

If we take the year 2000 as our starting point, we can say today that neurolaw has come of age. In the meantime, not only legal scholars have dealt extensively with the topics of neuroscience and law, as the afore-mentioned new edition of the almost thousand-page textbook shows best (Jones et al., 2022). As we have seen, it is also becoming more normal for courts to deal with the results of brain imaging and similar techniques. According to the results of my analysis, these possibilities fit into the existing normative order and do not overturn it, as has been claimed several times since the nineteenth century, especially in the free will debate.

In this chapter, I briefly summarize the main findings of the book. I then make a pragmatic proposal on the relationship between brain and behavior in normative contexts. Finally, I refer to open questions and future developments for neurolaw.

© The Author(s) 2025
S. Schleim, *Brain Development and the Law*, Palgrave Studies in Law, Neuroscience, and Human Behavior,
https://doi.org/10.1007/978-3-031-72362-9_5

5.1   GENERAL SUMMARY

We have not only traced the development phases of the human brain, but also of neurolaw in this book: In Chapters 1 and 3, we dealt with the emergence and substantive differentiation of the field; its maturation, so to speak. In Chapters 3 and 4, specific questions about criminal responsibility and the use of psychological and neurobiological knowledge were analyzed in detail, particularly with regard to human development. These investigations led us from "broken brains" and "teenage brains" to "(ir)responsible" or "dangerous brains" and finally even to alleged "cannabis brains." I write these terms all in quotation marks because, according to my point of view, brains and bodies must always be seen in their social and cultural contexts. More than once in the book, centuries-old findings played an important role as well.

In my opinion, however, such contextual factors are not the only reason why terms such as "dangerous brain" should at best be understood figuratively (e.g. Schleim, 2012). They should also be avoided for empirical scientific reasons: As I detailed elsewhere, 227 forms of depression can be distinguished based on the criteria of present the DSM diagnostic manual alone; for ADHD, the number is as high as 116,220 forms (Schleim, 2022a, 2023). Yet, researchers still use different methods to investigate depression, which was previously understood as "black bile disorder" (melancholia), with up to 52 symptoms instead of the only nine from the DSM (Fried, 2017).

In other words, there is no concrete "thing" depression or ADHD in the brains of those affected that could be detected using genetic or neurobiological methods. The heterogeneous, complex and processual nature of such phenomena makes it impossible to reduce them to simple biological structures and functions. Accordingly, Lombroso's project of a biological criminology was just as doomed to failure as phrenology. At best, one can expect biological characteristics with a certain predictive value, to which I will return in the last section of this chapter.

In Chapter 2, we learned about different ways of conceptualizing and differentiating human development from a psychological and neurobiological perspective. It was important to note that almost all changes occur gradually, with rare exceptions such as the first time of menstruation. From a medical and scientific perspective, terms such as "puberty" and "adolescence" are used to distinguish between different phases. However, these boundaries are also shifting due to the interaction of biological, psychological and social processes, and different understandings of adolescence, emerging adulthood and adulthood are used in parallel.

This, as well as the fact that many such terms are sometimes social constructs in a weaker, sometimes in a stronger sense, makes the distinction neither completely arbitrary nor useless. Rather, it illustrates the complexity of the topic and the different medical, social and scientific purposes that play a role here. With this result, it was already clear in principle that the categorical boundaries of the law discussed in Chapter 4 cannot fit the scientific models of development. We will shortly return to this point one last time.

First, in Chapter 3, we looked at the normative foundations of law and morality. The journey back to the nineteenth century illustrated the recurring dynamics of such debates: More than once, the idea suggested itself that statements about the alleged revolution of criminal law through scientific discoveries were more a strategy to attract attention than a careful scientific or philosophical argument.

On the one hand, the "neurorevolution" thesis presupposed a libertarian concept of free will, which even in today's philosophy is only held by a small minority (e.g. Roskies, 2006; Schleim, 2024a). On the other hand, the interpretation of recent experimental data against the possibility of conscious will turned out to be questionable. As we have seen, the criteria for criminal responsibility in particular are psycho-behavioral (e.g. Dressler, 2015; Morse, 2023). In short, they are based on the knowledge of right and wrong and the ability to act on this knowledge.

Now it is an empirical fact that we humans differ in characteristics such as intelligence or impulse control due to biological, psychological, social and situational factors. The law is able to distinguish between people with mental disabilities, psychological-psychiatric disorders or neurological diseases, those under the influence of drugs, in a serious moral dilemma, in a coercive situation or after the controlled planning of a crime. And such differences are regularly taken into account when determining responsibility, culpability and sentencing.

Neurological findings can be considered just as much as psychological examinations or witness statements. Anyone who wants to undermine this established useful, coherent and meaningful practice is not only up against everyday psychology. After all, psychological science also describes such differences between people and under different circumstances. Here, permanent characteristics tend to be assigned to the construct "personality" and spontaneous characteristics to the situation (e.g. Kuper et al., 2023). We have seen that even an offender with a large brain tumor, to take a frequently cited case, can be minimally rational, in the sense that he first hides evidence and then seeks help on his own (Burns & Swerdlow, 2003). Further research is needed to determine the extent to which pathological brain changes cause certain criminal behavior in individual cases.

Finally, in Chapter 4, more recent studies on psychological and brain development were related to criminal responsibility in particular. Such findings have played a role in specific legislative initiatives or Supreme Court decisions in the Netherlands and the USA. As we have seen repeatedly, however, three fundamental problems remain when drawing a concrete age limit: the variability within an age group, the gradual transition of development and the lack of practical relevance of neurobiological tests. But the latter might soon change due to the new studies on brain maturation now available, as described in Chapter 2.

Group and Individual

If a brain scan could be used to assess the *individual* development of an offender with a brain scan, this would undermine the *categorical* boundaries. Or why should a 17-year-old be judged in such a way, but not a 20- or 23-year-old? Psychology and neurobiology know no categorical boundaries here. Mental disability, for example, is also assessed on an individual basis. From a scientific point of view, however, following the sorites paradox known since antiquity—from how many grains of sand is a heap a heap (see Sainsbury, 2009)—*all* categorical boundaries could disappear: If someone is not fully responsible one day before their 18th birthday, then, based on their psychological/neurobiological development, they do not suddenly become so on their 18th birthday. The same applies to the day after their birthday, ad infinitum. In the end, no one would be fully responsible. This also demonstrates that such legal distinctions are based on (categorical) *normative* decisions, not (gradual and variable) *scientific* data.

In a constitutional state, laws must be general. Categorical distinctions based on a generally known factor such as age are easy to implement in practice. The criminal justice systems in Germany (up to and including the age of 21) and the Netherlands (up to and including the age of 22) know transitional phases in which *either* juvenile *or* adult criminal law can be applied, albeit with a different emphasis. The USA is a federal state with numerous jurisdictions with their own rules and practices. These laws allow for a certain degree of individual differentiation, albeit only for a particular age group.

The US Supreme Court case law discussed here also serves to curb arbitrariness in the imposition of maximum sentences, at least for juveniles (Banner, 2002; Denno, 2017). Even though psychological and neurobiological studies have informed authoritative decisions since 2005, the prospect of success of various ongoing initiatives to further raise the age limit is uncertain: This is due not only to the aforementioned fundamental problems of the scientific perspective, but also to the politically influenced majority in the court.

The argument that an increase would reduce racial discrimination (e.g. Haney et al., 2022) could be countered by arguing that it would be preferable to tackle general discrimination in society based on ethnic and social origin, education, gender and sexual orientation. In general, an exclusive focus on the brain threatens to overshadow psychosocial and contextual factors in the explanation of crime (Haney, 2020; Jalava et al., 2015; Schleim, 2024b). In this sense, it is somewhat ironic that the Dutch "criminal neurolaw"—with a view to preventing further crime—works primarily in that it keeps more offenders in their stable structures of home, work and social relationships (Prop et al., 2021). The *rationale* was neuroscientific, but the *intervention* is psychosocial.

In addition, the discussion of recent legislative initiatives such as the "cannabis brain" in Germany suggests that in political debates today, the desire to raise age limits may be given scientific weight simply by claiming that brain development has not yet been completed. As we saw in Chapter 2, this argument is somewhat tautological, since brain development is never fully completed. After a long discussion about whether neuroscientific information makes scientific explanations appear more credible, a new meta-analysis now confirms this suspicion (Bennett & McLaughlin, 2024). Yet, the effect is small. This may simply be mediated by the fact that neuroscience has been heavily emphasized in the media in recent decades, including reports of (alleged) breakthroughs (e.g. O'Connor et al., 2012; Racine et al., 2010). The examples in this book should have made it clear that it is nevertheless important to critically examine the plausibility and consistency of such findings.

The calls for higher age limits could be countered with the knowledge "against adolescence" from Chapter 2 and the critical objections of conservative judges on the US Supreme Court that most adolescents and young adults behave responsibly most of the time. Moreover, as we have seen, the reference to the continuous development of a plastic organ like the brain is trivially true. But if the societal trend of treating young people as adolescents for longer and longer periods of time prevails, this could lead to major legislative shifts. After all, there have always been such changes in the past. For our purposes, the only question that remains is what can be said today about the connection between brain and behavior from a pragmatic point of view.

5.2   A Pragmatic Approach

In the preface to the book, I already referred to the view that locates the brain in a body and both in a social situation. In cognitive science, the importance of this way of thinking is increasingly recognized again today and called "4E cognition" (Varela et al., 2017). According to this, our perception, thinking, feeling, decision-making and actions are necessarily embodied, embedded in a situation, enacted with and extending to the objects in the world. The discussion of numerous examples in the book emphasized the need to consider brain activations, structures and damage in connection with the other factors mentioned.

Even when we were probably the first in the world to specifically examine lawyers in the brain scanner, the resulting brain images did not provide an explanation in themselves; rather, they were a neuroscientific finding that *required* an explanation (Schleim et al., 2011). We asked our participants to judge normative problems alternately from a moral or a legal perspective: For example, is it permissible to call soldiers murderers? We would have liked to see evidence in the brain corresponding to the self-assessment of legal experts that they were less emotionally involved than other academics. But we did not find it. Whether this reflected a subjective misjudgment on the part of the lawyers or our experiment was not suitable for explaining the effect neurobiologically required follow-up studies. Unlike other experiments on moral judgment, which received a lot of attention at the time, we had focused on realistic problems, not on dilemmas with gory details and life-or-death decisions. Our stimuli thus were intrinsically less likely to trigger the emotional reactions reported by other studies.

In fact, it is still disputed whether the localization of brain function is at all useful for explaining our mental life. We addressed this question back in 2011 at a symposium entitled "Imaging the Mind? Taking Stock a Decade After the 'Decade of the Brain'"[1] which was only recently taken up again at a symposium of the Cognitive Neuroscience Society entitled "The Brain is Complex: Have we Been Studying it all Wrong?" (see Noble et al., 2024). We remember that this topic already played a decisive role in the description of the consequences of Phineas Gages' accident in the mid-nineteenth century—and remains relevant today (Schleim, 2022b). Accordingly, there are very different competing views on how cognitive processes and our behavior can best be explained (see Fig. 5.1). However, there are practical contexts in which one cannot wait forever for an answer from scientists and philosophers. In areas such as clinical psychology, psychiatry or law, a pragmatic approach is therefore required.

In the last two chapters, it has been repeatedly supported by both psychologists and legal scholars alike that the criteria of law are almost all psycho-behavioral (e.g. Morse, 2023). In criminal law, too, it is not brains on their own that commit crimes, but people in certain situations. We discussed in Chapter 3 that characteristics in the world and not just in the body and brain of the perpetrator can distinguish murder from attempted murder, for example—with very significant consequences for guilt and sentencing (Dressler, 2015). The primacy of psychology and behavioral sciences seems only logical when one considers that in criminal law certain *psychological processes* in combination with certain *criminal acts* are prohibited—and not brain states. We could concur with Jonathan Shedler, professor of psychology at the University of California in San Francisco, when he claimed: "It's Time for Psychology to Lead, Not Follow."[2]

We can illustrate this somewhat more systematically using a 2 × 2 matrix: Let's assume that a question can be answered positively or negatively from a psychological and neuroscientific perspective. As a concrete example, we can imagine the presence of depression, which can also play a role in employment law contexts. There are then two congruent and two incongruent possibilities (Table 5.1). In practice, it is of course possible

[1] The program and some of the presentations are still available online, see: https://www.schleim.info/imaging/.

[2] https://www.psychologytoday.com/us/blog/psychologically-minded/201910/its-time-for-psychology-to-lead-not-follow.

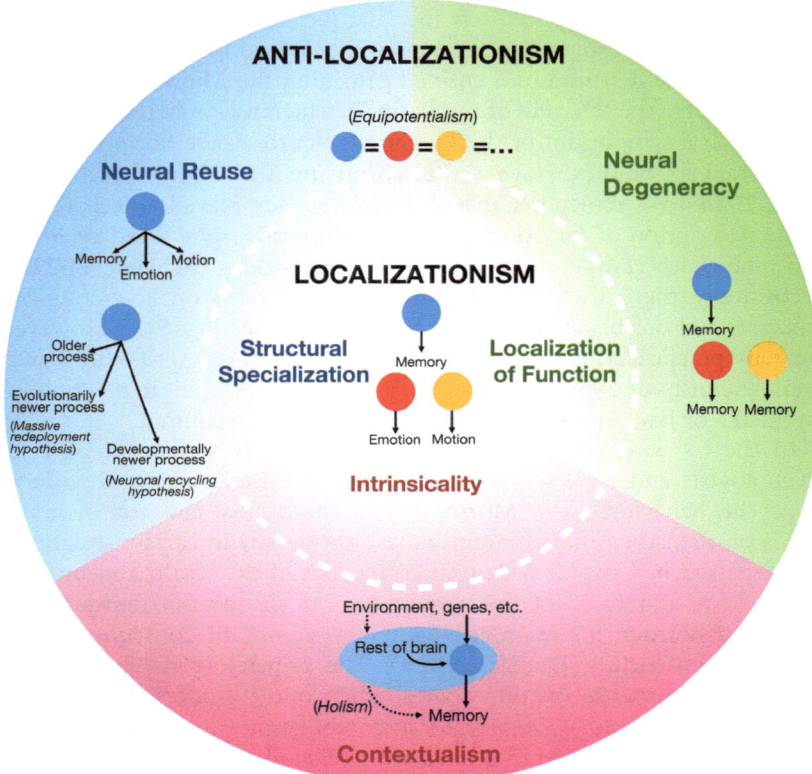

Fig. 5.1 The opposing poles shown here are localizationism in the middle and anti-localizationism on the outside. The three colors are used to differentiate the partial aspects of intrinsicality vs. contextualism (red), structural specialization vs. neural reuse (blue) and functional localization vs. neural degeneration (green). The gradual color transitions represent smooth transitions between the poles. For example, according to contextualism, the brain must be seen as a whole and in a specific environment in order to understand neuronal functions; according to neural reuse, new brain regions were not continuously created for new cognitive tasks in the course of evolution, but existing structures were reused in new ways. In both cases, results such as those from fMRI studies would be of limited explanatory value (*Source* Noble et al. [2024]. License: CC BY 4.0 [https://cre ativecommons.org/licenses/by/4.0/])

Table 5.1 The combination of two sources of information, in this case psychology and neuroscience, each with two possible answers provides four variants. The answers are congruent or incongruent for two of them. The latter are of particular interest for the theoretical discussion

	Neuroscience (+)	*Neuroscience (−)*
Psychology (+)	Congruent (+/+)	Incongruent (+/−)
Psychology (−)	Incongruent (−/+)	Congruent (−/−)

that a question cannot be answered from one side. The fact that other sources of information are then more important is trivial and is therefore not included in the table.

According to the currently widespread understanding of depression or "major depressive disorder," as the DSM calls it, various physiological, behavioral and psychological characteristics are characteristic of the disorder (e.g. Fried, 2017; Schleim, 2023). These include, for example, a lack of drive or motivation, depressed mood, more or less sleep, a change in weight that cannot be explained otherwise and thoughts of death.

Let's assume that someone would continue to function in their life as before, would not show any of the external characteristics and would not react conspicuously to questions about their state of mind. This would be the situation "psychology (−)," meaning that psychological evidence is *negative* about the presence of depression. We would then have to wonder very much how someone could still function normally if there were a positive finding "neuroscience (+)" (i.e. depression is present) from brain research. Mind you, as we noted in Chapter 1, such diagnostic markers would still have to be found in the first place.

Let's now take the opposite case: Someone would show the external and internal symptoms of depression, had trouble sleeping almost every night, had gained 5 kilograms in the last two months, would almost only lie in bed and report depressed feelings and thoughts when asked. However, this "psychology (+)" finding would be accompanied by a "neuroscience (−)" result. In the end, we would have to speculate about an error in the brain measurement—or about deception by the person in question. Incidentally, the discussion about malingering is as old as psychiatry and clinical psychology, precisely because of the more subjective nature of the symptoms and the lack of "objective" diagnostic markers.

This argument is in fact not only empirical but also theoretical; it also has something to do with the *meaning* of our language. To illustrate this point, we can look at even simpler examples, such as whether someone can catch a ball, drive a car, play chess or give meaningful answers in a particular language:

The fact that someone can drive a car means that they can regularly drive such a vehicle on the road from a starting point to a destination without causing accidents. This is a *fact in the world* that is established and can be independently verified by observation. If a neurologist were to claim that the person could not drive a car due to brain damage, we would perhaps have to assume remote control or—increasingly realistic today—intelligent assistance systems; or we would have to conclude that the neurologist is wrong. The example also illustrates the dependence of our psychological abilities on aids in the world according to the 4E perspective. To the extent that we use more and more external means, an intact nervous system becomes even less important for exercising certain abilities.

Practical Criteria

In Chapter 4, I already pointed out the curious discrepancy that the Dutch "criminal neurolaw" was essentially founded on the basis of *neuroscientific* studies, yet *psychosocial* criteria of the offender and/or the offense are decisive for its application. To clarify these, which are not specified in detail in the law, and as an alternative to the "intuition" of prosecutors, scientists have developed a number of criteria (Spanjaard et al., 2020). These were: (1) cognitive and adaptive skills (e.g. academic performance, deliberating on the consequences of one's actions), (2) social skills (e.g. verbal expression, friendships, ability to be influenced by others), (3) moral development (e.g. empathy, feelings of guilt) and (4) self-control (e.g. emotion regulation, risk-taking behavior).

Partly these *are in the world*, too, namely how someone lives and has relations with others; and partly they *are expressed in the world*, for example, how someone talks about a crime, whether they show remorse or not. If there were neurobiological tests for such or even better criteria, that would be extremely useful! In practice, the application of the rules of juvenile criminal law sometimes fails due to a lack of personnel for the preparation of psychosocial assessments (van der Laan et al., 2021). In addition, these could depend to a greater extent on the assessors themselves, just as psychological-psychiatric patients sometimes receive different diagnoses from different therapists or doctors (Schleim, 2023). With standardized neuroscientific procedures, the time and personnel required could probably be reduced considerably. However, the fact is that there are no corresponding biomarkers. This is precisely why it was so important to address the state of research in biological psychiatry and the problem of transferring neurobiological findings into practice in Chapter 1 (see also Schleim & Roiser, 2009).

When we discussed the possible link between a brain tumor and pedophilia in Chapter 3, the question ultimately remained whether the perpetrator was already sexually attracted to children and simply lost control of his behavior—or whether both this sexual inclination and the loss of control resulted from the neurological disease (Burns & Swerdlow, 2003). Similar to the improvement of the Dutch "criminal neurolaw," Italian researchers examined the literature for criteria to differentiate between permanent pedophilia and pedophilia "acquired" through an organic disease. On this basis, they evaluated 66 closed forensic cases. The three best criteria for identifying "acquired" pedophilia were that the perpetrator did not conceal his actions, confessed spontaneously or was of an advanced age (Ciani et al., 2019). The researchers also suggested these criteria as indications of whether an offender should be primarily investigated psychiatrically (permanent pedophilia) or neurologically (acquired).

Note that when we apply the Italians' clinical-empirical criteria to the case of the man with the brain tumor who sexually assaulted his stepdaughter, the evidence would rather speak against acquired (and thus in favor of permanent) pedophilic tendencies: He did conceal his actions until the victim sought refuge with her mother, he had not confessed his deeds spontaneously and he was only 40 years old at the time of the treatment (Burns & Swerdlow, 2003). Admittedly, the headline "Brain Tumor Causes Pedophilia" is catchy and makes neurolaw seem more relevant, which might explain that the case is still so frequently cited in the literature. I have criticized this interpretation for many years as too simple and neglecting important behavioral evidence (e.g. Schleim, 2012). Given the criteria by Andrea S. Ciani and colleagues (2019), I feel strengthened in my assessment that the tumor is more likely to have caused an impulse control disorder in addition to *pre-existing* pedophilic tendencies. This would also make an important difference considering that the man worked (and perhaps still works) as a school teacher.

Criteria as those of the Italian researchers can be further improved in terms of their usefulness and are coherent and meaningful. The latter results from their justification and embedding in previous research. Similarly, a court ruling also takes into account the legal situation, previous case law and practical possibilities and provides a comprehensible justification. This has worked so far and we have no reason to believe that it will be any different in the foreseeable future. This brings us to a final look ahead.

5.3   OUTLOOK

At the beginning of this chapter, I wrote that neurolaw has come of age. Representatives from various fields of law have now intensively examined the potential impact of neuroscientific procedures and will continue to do so (e.g. Jones et al., 2022).

Even after more than two unsuccessful centuries—or, if we think from antiquity, more than two millennia—researchers will continue to search for neural correlates of mental disorders, even if—with rare exceptions—they do not exist. The result on "broken brains" in Chapter 1 was clear in this regard and held up throughout the rest of the book. Perspectives on 4E cognition are, however, gradually playing a greater role in psychiatry (e.g. De Haan, 2020; Nielsen, 2023). This development could perhaps also extend to forensic research, combining sociological and perhaps even phenomenological models more closely with psychological and biological ones.

The extension of adolescence or emerging adulthood is taking place, as we saw in Chapter 2. Recently, there have been large collections of data on brain development from before birth to old age, which can serve as a reference in many areas. To my knowledge, however, these have not yet been used in forensic cases, which would tell us more about the individual application of such models.

The free will debate from Chapter 3 seems to have lost some of its momentum at the moment. Perhaps it will return in a new form in a few decades, as it has repeatedly done in the past. Other researchers and I have recently addressed how we humans can better deal with the unconscious influence on our decisions (Mudrik et al., 2022; Schleim, 2024a). These approaches also present free will as something that can be studied empirically and is sometimes more and sometimes less pronounced depending on the context and the person. This could ultimately provide finer distinctions that are important for the assessment of responsibility in practical contexts. For example, how far can manipulation in gambling go (e.g. Yücel et al., 2017) before a person can no longer legally consent?

The Dutch "criminal neurolaw" is currently being further evaluated and, as we saw in Chapter 4, various associations in the USA are seeking to increase the categorical age limit for maximum sentences to 21 years. One *can* argue that way from a scientific point of view—but we also saw fundamental problems here that need to be overcome. In my view, these efforts would become more convincing if a specific age limit for full criminal responsibility could be established instead of just claiming that the present one is too low. But in practice, these efforts are likely to fail for the time being because of the conservative orientation of the current US Supreme Court. Apart from that, actors in socio-political debates will now probably refer more frequently to the fact that the brain development is not yet completed when they want to raise an age limit.

Many of the examples in this book dealt with a more precise application of neuroscientific knowledge to the individual. But we are only gradually understanding more about the uniqueness of each brain (e.g. Jäncke & Valizadeh, 2022). From the "nine neurolaw predictions" of Morris B. Hoffman, a former district judge in the state of Colorado, five relate to individuals: From the perspective of 2018, he suggested that it should be possible to diagnose pain and legally relevant mental disorders using neuroscientific methods within the next ten years (Hoffman, 2018). I have just explained why I consider the latter to be very unlikely, if not impossible. Incidentally, the former director of the National Institute of Mental Health, Thomas Insel, was also wrong on all counts with such a prediction when he revived the idea of "broken brain circuits" as a model for mental disorders (Insel, 2010).

For the three other individual predictions, Hoffman considered a perspective of 50 years to be realistic: reliable lie detection, recognition of memories and the main topic of this book: the determination of a person's maturity. However, even more than 100 years after the first experiments with polygraphs, it is not so much due to the technical limitations but to conceptual problems that there are still no lie detectors for general use. People can give false information in many ways, sometimes in the belief that it is the truth, or complicate an investigation with psycho-behavioral countermeasures. To a certain extent, lies are also *in the world* and not just in our heads. Even the use of artificial intelligence has not yet led to a breakthrough in this area (see Suchotzki & Gamer, 2024). Thanks to the studies discussed in Chapter 2, individual maturity could already be determined neurobiologically today—but for the time being, these measurements do not appear to be any more informative than the established psychosocial methods. The large individual variance in brain structures and functions remains a fundamental problem here.

Interestingly, Hoffman did not mention an area that has already been intensively researched: the assessment of the dangerousness of an offender based on the prediction of the probability of recidivism (e.g. Zijlmans et al., 2021). The extent to which offenders could be forced to participate in such a measurement has already been investigated from a legal perspective (Ligthart, 2022). However, an offender also returns to a certain environment after a therapeutic measure and/or a prison sentence. If someone experienced enduring violence and deprivation, lost their social existence and carries a stigmatizing criminal record due to a prison sentence, the *prima facie* probability of new criminal behavior is higher (see also Scott et al., 2016). At least in part, this also means that the rate of recidivism lies *in the world*.

In the free will debate, the theoretical argument has been repeatedly put forward for over 100 years that the prediction of someone's behavior undermines its own preconditions if the person concerned is somehow treated differently or receives different information as a result (Schleim, 2024a). In addition, certain risk factors could be interpreted as an indication not only of particular dangerousness, but also of the need for special support. How society weighs up individual freedom against the general need for security in such cases is ultimately also an ethical, political and financial question.

The fact that neurolaw has come of age can also be seen in critical voices from the scientific community regarding such efforts: For example, researchers have shown that previous publications on the alleged "psychopathic brain" have various methodological flaws (Jalava et al., 2021, 2023). This question already plays a role today in the assessment of dangerousness, but also in the demand for reduced sentences. The researchers made the understandable suggestion that, when using such scientific results, courts should also consider the probability of error of a study and whether it has been independently replicated.

They even recently established a concrete decision tree for the admission of such findings in the courtroom: Whether (1) there are at least two independent studies relevant to the defendant's case in support of a certain conclusion, (2) these studies are adequately powered and (3) all relevant studies, thus also possibly contradicting evidence, have been disclosed (Jalava et al., 2023). Although these researchers identified and analyzed 64 scientific studies on the "psychopathic brain," they concluded that the available evidence base does *not* meet these eligibility criteria, which is both sobering and disturbing. As we discussed in Chapter 3, such studies have had—and continue to have—an impact on criminal trials in many cases.

In this book, we have focused primarily on the brain development of adolescents and young adults. Similarly, the aging brain and ultimately brain death also raise important legal and moral questions (e.g. Chandler, 2023). At the very beginning of life and the right to abortion, neuroscience recently played a role in the question of whether fetuses already feel pain (Salomons & Iannetti, 2022). However, it was not only the technological possibilities, but also the meaning of "pain" that was of crucial importance.

If the analyses presented here are not entirely wrong, "psycho-law" will remain more authoritative than neurolaw for the time being. This is not only due to our technological possibilities, but also the *meaning* of the terms we use to make sense of human behavior. Accordingly, there are established disciplines such as forensic and legal psychology, which can draw on neuroscientific methods if necessary. In any case, both the research and the discussion will continue and deliver many more interesting results.

REFERENCES

Banner, S. (2002). *The death penalty: An American history*. Harvard University Press.

Bennett, E. M., & McLaughlin, P. J. (2024). Neuroscience explanations really do satisfy: A systematic review and meta-analysis of the seductive allure of neuroscience. *Public Understanding of Science, 33*(3), 290–307.

Burns, J. M., & Swerdlow, R. H. (2003). Right orbitofrontal tumor with pedophilia symptom and constructional apraxia sign. *Archives of Neurology, 60*(3), 437–440.

Chandler, J. A. (2023). Cultural neuroethics in practice—Human rights law and brain death. In M. Farisco (Ed.), *Neuroethics and cultural diversity* (pp. 271–285). ISTE.

Ciani, A. S. C., Scarpazza, C., Covelli, V., & Battaglia, U. (2019). Profiling acquired pedophilic behavior: Retrospective analysis of 66 Italian forensic cases of pedophilia. *International Journal of Law and Psychiatry, 67*, 101508.

De Haan, S. (2020). *Enactive psychiatry*. Cambridge University Press.

Denno, D. W. (2017). Courting abolition. *Harvard Law Review, 130*(7), 1827–1876.

Dressler, J. (2015). *Understanding criminal law* (7th ed.). Matthew Bender & Company.

Fried, E. I. (2017). The 52 symptoms of major depression: Lack of content overlap among seven common depression scales. *Journal of Affective Disorders, 208*, 191–197.

Haney, C. (2020). *Criminality in context: The psychological foundations of criminal justice reform*. American Psychological Association.

Haney, C., Baumgartner, F. R., & Steele, K. (2022). Roper and race: The nature and effects of death penalty exclusions for juveniles and the "late adolescent class." *Journal of Pediatric Neuropsychology, 8*(4), 168–177.

Hoffman, M. B. (2018). Nine neurolaw predictions. *New Criminal Law Review, 21*, 212.

Insel, T. R. (2010). Faulty circuits. *Scientific American, 302*(4), 44–52.

Jalava, J., Griffiths, S., Larsen, R. R., & Alcott, B. E. (2021). Is the psychopathic brain an artifact of coding bias? A systematic review. *Frontiers in Psychology, 12*, 654336.

Jalava, J., Griffiths, S., & Larsen, R. R. (2023). How to keep unreproducible neuroimaging evidence out of court: A case study in fMRI and psychopathy. *Psychology, Public Policy, and Law, 29*(1), 1.

Jalava, J., Griffiths, S., & Maraun, M. (2015). *The myth of the born criminal: Psychopathy, neurobiology, and the creation of the modern degenerate*. University of Toronto Press.

Jäncke, L., & Valizadeh, S. A. (2022). Identification of individual subjects based on neuroanatomical measures obtained 7 years earlier. *European Journal of Neuroscience, 56*(5), 4642–4652.

Jones, O. D., Schall, J. D., & Shen, F. X. (Eds.). (2022). *Law and neuroscience* (2nd ed.). Aspen Publishing.

Kuper, N., von Garrel, A. S., Wiernik, B. M., Phan, L. V., Modersitzki, N., & Rauthmann, J. F. (2023). Distinguishing four types of Person × Situation interactions: An integrative framework and empirical examination. *Journal of Personality and Social Psychology, 126*, 282–311.

Ligthart, S. (2022). *Coercive brain-reading in criminal justice: An analysis of European human rights law*. Cambridge University Press.

Morse, S. J. (2023). Neurolaw: Challenges and limits. *Handbook of Clinical Neurology, 197*, 235–250.

Mudrik, L., Arie, I. G., Amir, Y., Shir, Y., Hieronymi, P., Maoz, U., et al. (2022). Free will without consciousness? *Trends in Cognitive Sciences, 26*(7), 555–566.

Nielsen, K. (2023). *Embodied, embedded, and enactive psychopathology reimagining mental disorder.* Palgrave Macmillan.

Noble, S., Curtiss, J., Pessoa, L., & Scheinost, D. (2024). The tip of the iceberg: A call to embrace anti-localizationism in human neuroscience research. *Imaging Neuroscience, 2,* 1–10.

O'Connor, C., Rees, G., & Joffe, H. (2012). Neuroscience in the public sphere. *Neuron, 74*(2), 220–226.

Prop, L. J. C., Beerthuizen, M. G. J. C., & Van der Laan, A. M. (2021). *Adolescentenstrafrecht: Effecten van de toepassing van het jeugdstrafrecht bij jongvolwassenen op resocialisatie en recidive.* Wetenschappelijk Onderzoek- en Datacentrum.

Racine, E., Waldman, S., Rosenberg, J., & Illes, J. (2010). Contemporary neuroscience in the media. *Social Science & Medicine, 71,* 725–733.

Roskies, A. (2006). Neuroscientific challenges to free will and responsibility. *Trends in Cognitive Sciences, 10*(9), 419–423.

Sainsbury, R. M. (2009). *Paradoxes* (3rd ed.). Cambridge University Press.

Salomons, T. V., & Iannetti, G. D. (2022). Fetal pain and its relevance to abortion policy. *Nature Neuroscience, 25*(11), 1396–1398.

Schleim, S. (2012). Brains in context in the neurolaw debate: The examples of free will and "dangerous" brains. *International Journal of Law and Psychiatry, 35*(2), 104–111.

Schleim, S. (2022a). Why mental disorders are brain disorders. And why they are not: ADHD and the challenges of heterogeneity and reification. *Frontiers in Psychiatry, 13,* 943049.

Schleim, S. (2022b). Neuroscience education begins with good science: Communication about Phineas Gage (1823–1860), one of neurology's most-famous patients, in scientific articles. *Frontiers in Human Neuroscience, 16,* 734174.

Schleim, S. (2023). *Mental health and enhancement: Substance use and its social implications.* Palgrave Macmillan.

Schleim, S. (2024a). *Science and free will: Neurophilosophical controversies and what it means to be human.* Springer.

Schleim, S. (2024b). De bijzondere rol van de neurowetenschappen in het Nederlandse strafrecht. *Podium voor Bio-ethiek, 31*(1), 18–24.

Schleim, S., & Roiser, J. P. (2009). FMRI in translation: The challenges facing real-world applications. *Frontiers in Human Neuroscience, 3,* 845.

Schleim, S., Spranger, T. M., Erk, S., & Walter, H. (2011). From moral to legal judgment: The influence of normative context in lawyers and other academics. *Social Cognitive and Affective Neuroscience, 6*(1), 48–57.

Scott, E. S., Bonnie, R. J., & Steinberg, L. (2016). Young adulthood as a transitional legal category: Science, social change, and justice policy. *Fordham Law Review, 85*, 641–666.

Spanjaard, H. J. M., Filé, L. L., Noom, M. J., & Buysse, W. H. (2020). *Achterlopende ontwikkeling: Het begrip 'onvoltooide ontwikkeling' in de toepassing van het adolescentenstrafrecht*. Universiteit van Amsterdam.

Suchotzki, K., & Gamer, M. (2024). Detecting deception with artificial intelligence: promises and perils. *Trends in Cognitive Sciences, 28*, 481–483.

van der Laan, A. M., Beerthuizen, M. G., & Barendregt, C. S. (2021). Juvenile sanctions for young adults in the Netherlands: A developmental perspective. *European Journal of Criminology, 18*(4), 526–546.

Varela, F. J., Thompson, E., & Rosch, E. (2017). *The embodied mind, revised edition: Cognitive science and human experience*. MIT press.

Yücel, M., Carter, A., Allen, A. R., Balleine, B., Clark, L., Dowling, N. A., et al. (2017). Neuroscience in gambling policy and treatment: An interdisciplinary perspective. *The Lancet Psychiatry, 4*(6), 501–506.

Zijlmans, J., Marhe, R., Bevaart, F., Van Duin, L., Luijks, M. J. A., Franken, I., et al. (2021). The predictive value of neurobiological measures for recidivism in delinquent male young adults. *Journal of Psychiatry and Neuroscience, 46*(2), E271–E280.

INDEX

© The Author(s) 2025
S. Schleim, *Brain Development and the Law*, Palgrave Studies in Law,
Neuroscience, and Human Behavior,
https://doi.org/10.1007/978-3-031-72362-9